CAMBRIDGE BIOLOGICAL STUDIES
General Editor: C. H. WADDINGTON

THEORIES OF SENSATION

THEORIES OF SENSATION

by

A. F. RAWDON-SMITH

M.A. (Cantab.), Ph.D. (Cantab.)

CAMBRIDGE

AT THE UNIVERSITY PRESS

1938

CAMBRIDGE
UNIVERSITY PRESS

University Printing House, Cambridge CB2 8BS, United Kingdom

Published in the United States of America by Cambridge University Press, New York

Cambridge University Press is part of the University of Cambridge.

It furthers the University's mission by disseminating knowledge in the pursuit of education, learning and research at the highest international levels of excellence.

www.cambridge.org
Information on this title: www.cambridge.org/9781107678484

© Cambridge University Press 1938

First published 1938
First paperback edition 2014

A catalogue record for this publication is available from the British Library

ISBN 978-1-107-67848-4 Paperback

To

PROFESSOR E. D. ADRIAN

whose work has so greatly influenced our present
Theories of Sensation

CONTENTS

ACKNOWLEDGEMENT

This book owes very much to the many people with whom I have been privileged to discuss its contents. From all those with whom I have from time to time collaborated I have learnt much more than I am likely to be able to repay, and much of what they have taught me is incorporated here. It would be difficult, if indeed desirable, for me to make specific reference to all these many kindnesses, and with some exceptions this general acknowledgement must suffice. I am conscious of particular debts, however, to which I must make special reference.

The manuscript of Section I has benefited very greatly from the critical inspection to which my friend Mr K. J. W. Craik has subjected it. Much of this section I owe to his help, though for such errors of commission and omission as remain he is in no way responsible. The index has been entirely prepared by my friend and collaborator Mr R. S. Sturdy, to whom the presence of this valuable feature is altogether due.

It would be impossible for me to overestimate the importance of the many opportunities which I have been afforded of talking over all manner of problems with Professor F. C. Bartlett and with Professor E. D. Adrian, and with other members of their departments. I am grateful too for the interesting discussions which I had with Dr Hallowell Davis, of Harvard University, during his visit to England in the summer of 1935. But it would be quite improper for me to ascribe to any of these the opinions which I have expounded here. For the preparation of the present exposition, both of my own ideas and of what I have understood of others', I must be held wholly responsible.

To my wife I am particularly indebted, on the one hand for her encouragement during the moments of depression inevitable when writing a book, on the other for her discouragement of my more violent abuse of the English language.

x ACKNOWLEDGEMENT

To Dr C. S. Hallpike I am grateful for his permission to
reproduce Figs. 14 and 18, here published for the first time.
Dr W. D. Wright has been kind enough to allow me to use
data of his for the construction of the graphs shown in Figs. 7
and 12. Dr S. S. Stevens and Dr L. M. Hurvich, of Harvard
University, have read this book in proof, and have helped me
with many valuable suggestions.

To the following, I am indebted for their permission to repro-
duce various figures: Messrs Longmans, Green, Fig. 1; The
Rockefeller Institute of Medical Research, Figs. 2 and 3; Clark
University Press, Figs. 4 and 11; The Wistar Institute of
Anatomy, Figs. 5 and 6; The Optical Society of London, Fig. 7;
Cambridge University Press, Fig. 8; Johann Ambrosius Barth,
Fig. 9; Messrs J. and A. Churchill, Fig. 13; and the American
Physiological Society, Fig. 16.

Fig. 10 is reprinted by permission from *Psychology* by Boring,
Langfeld and Weld, published by John Wiley and Sons, Inc.

A. F. RAWDON-SMITH

Cambridge, England; 1937

Cambridge, Massachusetts; 1938

INTRODUCTION

In common with a majority of the higher animals, man is ordinarily held to possess five senses. Of these, the first four, namely sight, hearing, taste and smell, are to some extent unlike the fifth, which is in reality a group of senses, mediating our sensibility to touch, pressure, temperature and pain. Thus, because their functions are accomplished by specific sense organs, the eyes, ears, taste-buds and nose, the first four are usually known as the special senses; whilst the remaining group, known as the general senses, are so called because the corresponding sensations can be aroused by stimulation of almost any part of the external surface of the body, and indeed of a large number of internal organs in addition. Thus, muscles and tendons are sensitive to pressure or tension, to painful stimuli, and to some extent to temperature, and the peritoneum, which surrounds the intestines, is sensitive at least to pressure and pain.

In view of their importance in enabling his adaptation to and control of his environment it is not surprising that man has long speculated over the mechanism of operation of his senses. Inevitably, such speculation led to the elaboration of theories of sensory operation. It is a canon of the scientific method that speculation, unalloyed with results of the more direct experimental method, may, however, as often mislead as assist. The field of sensory physiology gives one no cause to dispute this. Here, as elsewhere, the more recent experimental technique has always modified and often clarified the early theoretical position.[1] Unhappily, however, clarification has not been the invariable result of its application. Our knowledge of the processes of vision and audition has, it is true, advanced enormously. The present theoretical position is consequently of the greatest interest. The general senses, have, however, up to the present

[1] Thus we read of the Greek philosopher who, by purely speculative methods, deduced that in the goat the ear is the organ of respiration, a view which later experimentalists have seldom espoused!

suffered the opposite fate. Thus the discovery by the early histologists that the nerve endings of the skin could be divided into a number of morphologically dissimilar groups led to the early elaboration of the plausible hypothesis that the different sensations resulting from their stimulation could be correlated with differences of structure. More recently, however, the justification for this view has largely disappeared, for not only does it seem that many types of end-organ intermediate between the recognised groups can in fact be found but, worse, it would appear that certain end-organs may yield a variety of sensations under appropriate conditions of stimulation. It would be safe to say that no theory of general sensibility has yet been proposed which is able to describe the available data.

Returning to the organs of special sense, we may note, in the fields of olfaction and gustation, not that the theoretical position is insecure, but simply that no theoretical position exists. The simple statement that our sense of smell depends upon stimulation of the olfactory epithelial end-organs, or that taste results from stimulation of the taste-buds of the papillae of the tongue, is all that, with confidence, can be said. The anatomical inaccessibility of these organs in the live animal, combined with the almost complete absence of simultaneous discrimination in the human subject, has resulted in the fields of physiological and psychological investigation respectively being virtually barren. This is not to say that there is not a considerable literature available here; that the varied data collected have suggested no satisfactory theory even as to the mechanism of stimulation of these organs is, however, only too evident.

In this book I shall mainly be concerned with theories of sensation—such experimental data as are recorded should find a place here only so far as they are relevant to the theories under discussion, for I wish to be concerned more with the mechanism whereby our sensations arise than with what our sense organs *can* do. In the fields of general sensibility, olfaction, and gustation the theoretical position, as I have said, is either hopelessly insecure or entirely absent. Indeed, little more of theoretical interest could well be added to the brief treatment contained in this introduction. Accordingly, in this book I shall be concerned only with theories of vision and audition, and with the data of these senses as they are relevant. However, to those who wish to extend their knowledge to cover the work which has been done on the remaining sense organs, it may perhaps be helpful to point to the existence of von Skramlik's classical monograph on taste and smell,[1] which suffers only from the limitation that it is inevitably not altogether up to date, and to Nafe's able summary of the position of general sensibility, published in 1934.[2] To these the reader interested in the other senses may safely be directed.

In the present work I shall attempt to describe, in the greatest detail consistent with its length, how it is now thought that our eyes and ears work, and, as far as possible, why these views are held in preference to others. As our discussion of the matter advances we shall see that, on a large number of points, opinions differ—a state of affairs inevitable in a subject at once so difficult and, so far, so little investigated. In such circumstances I shall attempt to give both sides of the argument, though I shall also try to indicate which appears to me to be the more likely view. On some points this will, I fear, indicate merely some personal preference, whilst on still others it will not be possible to differentiate the conflicting theories. No apology need be made for this, for on very many points the evidence has yet to accumulate. However, we are already some way towards understanding at least the earlier and simpler processes leading to what we call a *sensation*.

It is in the hope of interesting other biologists in this intriguing and fundamental field that this book is written.

[1] *Handbuch der Physiologie der niederen Sinne*, by Emil von Skramlik. 1. Band: *Die Physiologie des Geruch- und Geschmacksinnes*. Leipzig: Georg Thieme, 1926.
[2] "The Pressure, Pain and Temperature Senses", by John Paul Nafe. Pp. 1037 to 1087 in Murchison's *Handbook of General Experimental Psychology*. Worcester, Massachusetts: Clark University Press, 1934.

SECTION I
VISION

CHAPTER 1

The Formation of a Retinal Image

Anatomically,[1] the human eye consists of an approximately spherical, hollow object whose external diameter is ordinarily about 2·3 cm. It is built up of several layers, and it is convenient to describe these in order starting with those comprising the external surface and proceeding inwards. A horizontal section of the eye (Fig. 1) shows, on the exterior surface, a tough outer coat known as the *sclera* or sclerotic; or popularly as the "white of the eye". Anteriorly, the sclerotic is continuous with the transparent *cornea*. The latter is somewhat more highly curved than is the sclerotic and it is here that the shape of the eye as a whole departs to the greatest degree from the truly spherical.

Within the sclerotic is the *choroid*, a layer containing the great majority of the blood vessels of the eye and of very dark colour. The innermost layer is the retina, which contains the photosensitive elements, and whose structure will be discussed more fully later. Immediately behind the corneoscleral junction the retina and choroid layers are modified to comprise the *ciliary body*, which contains, *inter alia*, the ciliary muscles and their blood supply, and from which, on the anterior surface, the *iris* arises. The ciliary muscles are of two distinct types. The meridional fibres, or *Brücke's muscle*, are those nearest the external surface. They run approximately tangentially to the sclerotic, on its inward face. Between these and the *suspensory ligament* of the lens lie the circular fibres, or *Müller's muscle*. In Fig. 1 these are cut in transverse section, and it will be seen, therefore, that they run at right angles to the meridional fibres, and constitute a sphincter round the lens.

The iris itself comprises two layers, one continuous with the retina and the other with the choroid.

---

[1] It is scarcely necessary to point out that this anatomical description makes no attempt at being comprehensive. It is inserted for the sake of completeness, as constituting the minimum of knowledge required for an understanding of the visual processes to be described later. More detailed accounts will be found in most anatomical text-books, of which that in Quain's *Anatomy* (1909) is especially recommended.

Attached to the ciliary body just posterior to the iris is the *suspensory ligament* of the *crystalline lens*. The latter separate the remaining contents of the interior of the eyeball, the *aqueous* and *vitreous humours*.

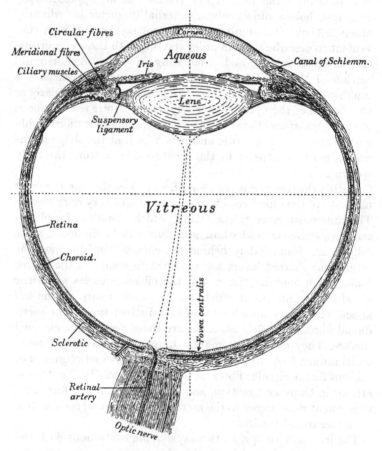

Fig. 1. Diagram of the human eye (after Quain).

The first essential condition for the production of the visual sensation is the formation of an inverted image of the external object field upon the retina. The chief refracting surface enabling this is the outer surface of the cornea. The crystalline lens proper, surrounded as it is by fluids whose refractive indices are little less than its own, makes but a secondary contribution to the total refraction. Its importance lies, however, in the fact that its radius of curvature, and hence its refractive power, may be altered, thus permitting the process of *accommodation* whereby objects anywhere in the visual field may be brought into sharp focus on the retina, provided that their distance (for the normal eye) lies between the near point[1] and infinity.

Our present view as to the mechanism whereby this is accomplished in the human eye is due to Helmholtz, and is ordinarily known as Helmholtz' Theory of Accommodation (Helmholtz, 1924, pp. 143–71). It is based on the fact that if the tension exerted on the circumference of the lens by the suspensory ligament is released, as by excising the lens, it will assume a shape more nearly spherical than in the presence of this tension. Consideration of the mechanics of the ciliary body reveals that, when the ciliary muscles contract, tension on the suspensory ligament will be partially relaxed. Contraction of the fibres of Müller's muscle will clearly reduce the diameter of the ring which they form, whilst contraction of the meridional fibres (Brücke's muscle) will also release the tension on the ligament, the posterior ends of these fibres being drawn forwards and inwards towards the corneoscleral junction. In virtue of its own elasticity the radius of curvature of the lens will now decrease, thereby bringing objects nearer to the eye into sharp focus on the retina. The anterior surface of the lens capsule is not, however, uniformly thick. Those parts of the anterior surface nearer the suspensory ligament are thicker than are those nearer the centre or anterior pole. The decreased radius of curvature on accommodation, therefore, takes effect rather in the centre than at the sides. Since only the centre becomes more spherical (the thickened sides remaining approximately straight) it is necessary, if accommodation for objects nearer the eye is not to result in serious distortion of the retinal image, for the diameter of the pupil to decrease, thereby covering up the less curved part of the lens where the lens capsule is thickest, and exposing only the

---

[1] The "near point" is simply defined as the point nearest to the eye at which an object may be brought into sharp focus.

highly spherical frontal surface. The reflex whereby this is
accomplished simultaneously with accommodation for near
objects is known as the Pupillary Accommodation Reflex. In
brief, then, accommodation for near objects causes the diameter
of the pupil to decrease, even though the total light impinging
upon the eye does not change. To this point we shall return
later.

The precise nature of the changes occurring to the lens on
accommodation may be quite readily demonstrated in the intact
human being. If a small source of light, for example the flame of
a candle, is brought near to the eye of such a subject, three
images of the flame may be seen in the eye (Helmholtz, 1924,
pp. 144–6). The brightest of these is uninverted and is produced
by the anterior corneal surface acting as a convex mirror. Less
bright than this, a second uninverted image will be seen, reflected
from the convex anterior lens surface. A third image, very dim
and inverted, may with care be observed in the concave posterior
surface of the lens. These three are known as *Sanson's images*,
and the changes they undergo during accommodation give a
considerable amount of information as to the nature of the
process itself. These changes may be summarised as follows. The
image from the cornea undergoes no change, wherever the eye is
accommodated from infinity to the near-point. That produced
by the posterior lens surface undergoes only a minor change,
showing that the radius of curvature of this surface is not
markedly altered. The uninverted anterior lens surface image,
however, is very much larger when the eye is accommodated for
distant objects than when it is focused on nearer ones. This
indicates that the major change during accommodation occurs
to the anterior lens surface; and, moreover, as the image here
becomes smaller when the eye accommodates for nearer objects
we may further specify that the nature of this change is such
that this surface decreases its radius of curvature.

The anatomical position of Brücke's muscle is such that it
might be expected that, during contraction of this muscle (that
is to say during accommodation for near objects), the whole lens
capsule might move forward. Direct measurement shows,
however, that this is not the case, for the posterior lens surface

actually moves very slightly backwards, whilst the anterior surface moves rather farther forwards.

It has been mentioned above that during accommodation the pupil contracts, a process known as the Pupillary Accommodation Reflex. If the illumination of the eye increases, a similar effect may be observed—the Pupillary Light Reflex. A third reflex may be observed if one eye is covered up whilst the other is strongly illuminated. The resultant contraction of the pupil of the unilluminated eye is known as the Consensual Pupillary Reflex.

The mechanism whereby these changes of pupillary diameter (which varies between 2 and 8 mm. for the normal subject) are accomplished is as follows. The iris possesses two muscles, one circular and ordinarily known as the *sphincter pupillae* and one whose fibres are radial, known as the *dilator pupillae*. Of these, the former exerts very much more effect than does the latter, and is supplied by the III or oculo-motor nerve. The *dilator pupillae* is supplied by fibres from the sympathetic nervous system. In the absence of impulses from the III nerve to the sphincter, the normal tone of the dilator causes the pupil to assume its maximum diameter. The antagonistic sphincter can, however, constrict the pupil despite the maintenance of dilator tone; the three pupillary reflexes mentioned above are mediated mainly by impulses in the III nerve, the sympathetic impulses then having little effect on pupillary diameter.

We may now briefly summarise what is known of the optical system of the eye as follows. An inverted image of the external object field is formed in the retina by refraction of the incident light rays at the anterior corneal surface and anterior and posterior surfaces of the crystalline lens. The radius of curvature of the lens surfaces, more particularly the outer of these, may be changed during the process of accommodation, thereby enabling the retinal image to be in sharp focus, irrespective, within limits, of the distance from the eye of the object producing the image. Accommodation for near objects produces a simultaneous constriction of the pupil, independently of any change in the magnitude of the illumination falling on the eye, thereby compensating optical defects of the near-accommodated lens. An increase in this illumination, to one or both eyes, will similarly decrease pupillary diameter, thus maintaining the brightness of the retinal image, during changes of illumination of the external object field, more nearly constant than would otherwise be the case. It has recently been shown (Stiles and Crawford, 1933), however, that the effect of the pupil is smaller than was previously believed; according to these workers it can rarely accomplish a change of retinal illumination of more than 3 : 1.

The retinal image cannot correctly be imagined simply as an inverted and reduced copy of the external object field. As is inevitable from the supposed nature of light,[1] a point source will produce, on the retina, an image consisting of a somewhat complicated diffraction pattern. This effect is, of course, an "interference" phenomenon, and cannot be considered as a defect of the eye. It would be expected that the size of the pattern produced would vary inversely as the aperture of the lens, and thus as the diameter of the pupil; thus its magnitude will increase as the pupillary diameter is reduced.

In addition, the eye suffers from various forms of aberration. Of these the more important are chromatic, whereby the constituent wave-lengths of white light are differentially treated with regard to magnification and focus by the optical system. These have been very fully treated by Hartridge (1918), and it will suffice here to point out that their effects will be such as still further to enlarge the size of the retinal image of a point source. There is good evidence for the supposition that these effects will vary directly with pupil diameter, and thus there would appear to be some degree of compensation between diffraction and the various forms of chromatic aberration, the size of the retinal image thus tending to be constant irrespective of pupil size (Hartridge, 1923).

The importance of these phenomena will be seen when it is realised that the eye's resolving power apparently depends to a great extent upon their existence. To this matter we shall, however, return later.

[1] The physical concept of light is undergoing some modification at the present time. Fortunately, however, visual phenomena can be adequately treated in terms of the early wave theory of Huyghens and Young. For this reason, we need not here be concerned with this difficult field of scientific thought.

CHAPTER 2

The Duality of the Retinal Process

The innermost layer of the eyeball comprises the retina, or photosensitive layer; its angular extent is, in the normal eye, c. 207°. Approximately in the retinal centre there is a small depression, known as the *fovea centralis*, whilst some 18° off centre, to the nasal side, the retina is perforated by the exiting fibres of the optic nerve, photosensitive elements being absent over this area. When an image is formed here, therefore, the object producing it cannot be seen; that part of the external object field to which such an image corresponds is hence known as the "blind spot". Microscopically, the retina may be shown to consist of three distinct layers (Quain, 1909, pp. 220-52). Proceeding from the choroid inwards, these are (1) the receptor cells, consisting of two ordinarily distinguishable types, the *rods* and *cones*, (2) the *bipolar neuron layer* occasionally subdivided into inner and outer molecular and nuclear layers, and (3) a layer comprising ganglion cells, with, further inwards still, their attached axons, the latter combining to form the optic nerve. On the outer surface ramifying branches of the retinal artery are also present.

The first of these layers contains, as has been said, two distinct types of cells, the photosensitive rods and cones. These cells are so named because of the shape of their terminal segments, which are further supposed to be the locus of sensory excitation. At the fovea, rod cells are believed to be absent over an area corresponding to a horizontal diameter of 1·7° (Kohlrausch, 1931, p. 1520), though at this point the cone cells are so closely packed together that in shape they more nearly resemble the rod cells farther out in the retina (Greef, in Quain, 1909). There is, how-ever, no corresponding functional similarity.

One fact instantly emerges from this anatomical description. It will be seen that a ray of light, before it can yield any sensation, must first pass between the branches of the artery ramifying on the retinal surface, thereafter through the ganglion cell, axon and bipolar neuron layers, before reaching the receptor layer at which stimulation takes place. It is, therefore, a matter of some considerable interest that, at the *fovea centralis*, that region of the retina mediating most distinct vision and highest acuity, the innermost layers of the retina, the so-called nuclear and molecular layers, are very much thinner than elsewhere. Moreover, though surrounded by blood vessels, the fovea itself has none, and the cone cells themselves, therefore, more closely approximate to the inner surface. In this way the obstruction offered to rays of light reaching the receptor layer at this point is minimised, though not altogether removed. This is of considerable importance in view of the fact that the eyes are invariably so moved that the image of any object undergoing close examination at normal illuminations falls upon the fovea, and, as has been said, it is here that the eye's acuity is greatest.

The fovea itself lies in the centre of the *macular plexus*, a region of the retina where acuity is higher than elsewhere, though not so high as at the fovea proper. Over the macular plexus the ganglion cell and outer molecular layers are considerably thicker than at more peripheral parts of the retina, and this region appears, moreover, to be especially well represented in the cerebral cortex. Corresponding in area and position with the macular plexus is a thin layer of yellow pigment (the *macula lutea*) whose function is not understood.

Important functional differences accompany the histological differentiation of the retinal photosensitive elements into rods and cones. In the human eye it is readily possible to study visual function which is solely cone mediated, in the 1·7° rod-free area centred at the fovea. As no part of the human retina is wholly cone-free, however (Troland, 1934), other means must be adopted in order to determine those functions which are exclusively rod mediated.

Our present knowledge of the functional differences of these two receptor types is based to some extent upon observations of the distribution of the two in species of dissimilar habits. Thus, nocturnal animals possess retinae largely composed of rods, whilst diurnal species ordinarily show a relatively greater proportion of cones. The generally accepted view that the two cell types mediate vision at different intensities of illumination—the cones at high and the rods at low intensities—is known as the Duplicity Theory of von Kries (1895, 1929), though it had been suggested in similar form as early as 1866 by Schultze. An adequate historical summary is given by Parsons (1924, pp. 215–24). A considerable amount of evidence in favour of this view has arisen from the study of so-called Visibility Curves, in which attempts are made to establish a relation between brightness and wave-length for homogeneous spectral stimuli. The most often employed technique is that in which perceived brightness is held constant for varying wave-lengths by making the requisite adjustment of physical intensity. Using a spectrum of reasonable overall brightness, to which, moreover, the test eye is adapted, the curve thus produced shows a minimum at a wave-length of 554 mμ., ascending on either side towards the spectral extremes (Gibson and Tyndall, 1923). If the overall test intensity is now considerably reduced, the minimum shifts to 511 mμ., the general form of this second curve being, however, substantially identical with that of the first (Hecht and Williams, 1922) (Fig. 2). It is convenient to term these two curves *photopic* and *scotopic* for the brighter and less bright respectively. In the curves shown the *reciprocals* of the energy values obtained by the technique described above have been plotted to give the so-called visibility coefficients.

The displacement of the region of maximum visibility down the wave-length scale as the brightness is decreased is responsible for the Purkinje Phenomenon, and certain related effects. Conventionally, this effect is described as a movement of the brightest region of the spectrum of white light from the "yellow" to the "green" wave-lengths, in consequence of a reduction in overall intensity. Further, during this process the long-wave end of the spectrum becomes considerably dimmer or even invisible, whilst the brightness of the short-wave end demonstrates a corresponding relative increase.

From the rods of the mammalian eye a substance known as *visual purple* (Boll, 1876) may be extracted by means of certain haemolytic agents (Kühne, 1879), with the production of a pink solution which is bleached if light is allowed to fall upon it. Trendelenburg (1904, 1911) has shown that a very close similarity exists between the curve connecting rate of bleaching of visual purple with wave-length and that showing relative brightness against wave-length (measured by the technique outlined above) for the scotopic eye (Fig. 3). This is accepted as strong evidence that the bleaching of visual purple is an im-

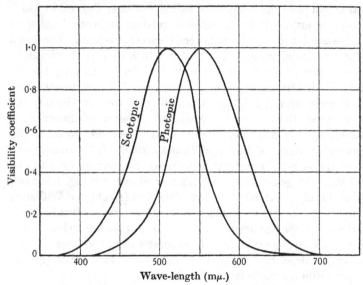

Fig. 2. Photopic and scotopic spectral visibility curves
(after Hecht and Williams).

portant stage in the production of rod vision. It should, however, be pointed out that the curves are not, in fact, identical, for the visibility curve is displaced some 7 mμ. towards the red end of the spectrum. It has been suggested that this displacement is due to the difference in medium, the visual purple being, on the one hand, in aqueous solution, and on the other in the highly refractive outer rod segments. This may well be the case. It is not, however, possible to predict either the direction or the magnitude of the shift, as Hecht (1934, 1937) has attempted to do, in terms of Kundt's rule (Kundt, 1878), as this rule is no longer accepted (Houstoun, 1932a, p. 93). Lythgoe (1937) has shown that the absorption curve is somewhat sensitive to the pH of the solution, and it seems legitimate to conclude that, allowing for this and for absorption by the intra-ocular media, the curves are in fact identical (Ludvigh, 1938).

The bleaching effect of visual purple is responsible for the so-called "optogram" in the intact mammalian eye. If such a dark-adapted eye is excised, and the image of a bright object is then allowed to fall upon the retina for some minutes, an in-

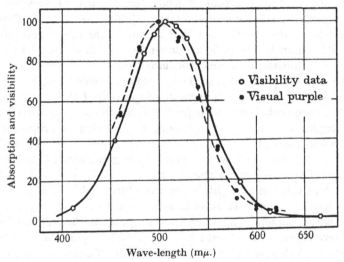

Fig. 3. The scotopic visibility curve for the human eye, and the spectral absorption curve of visual purple (after Hecht and Williams).

verted image of the object will be found upon the retina if the eye is plunged into alum and dissected. It is generally accepted that the bleaching of the rod-contained visual purple produces this effect.

No substance similarly responsible for cone-mediated vision has yet been identified. It has been suggested that visual purple, in rather more dilute form, is responsible here also (Weigert, 1921; Hecht and Williams, 1922), though Hecht (1934) has now abandoned this view. There would appear to be no valid evidence for it, for, although the substance or substances responsible may well be closely related to visual purple, the differences between the photopic and scotopic visibility curves are probably too large to be explained in terms of differences in concentration, though such an explanation has been attempted (Hecht and Williams, 1922).

Numerous other differences between rod-mediated and cone-mediated vision exist. It is well known, for example, that colour vision is almost solely a function of the cones, a supposition whose validity may be readily illustrated by reducing the intensity of the spectrum of white light which, at very low illuminations, will be seen as a colourless band of light. This effect is ascribed to the fact that the cones are very much less sensitive than are the rods, by which the colourless vision at the lowest illuminations is solely mediated.

Other differences between the two types of vision appear upon examination of the phenomena of adaptation of the eye. It is well known that, if the eye is protected from light its sensitivity undergoes a marked enhancement, a phenomenon ordinarily known as *Dark Adaptation*. The time which such a process will take for its completion will depend very largely upon the state of *Bright Adaptation* which the eye has previously achieved, and therefore upon the brightness of the object field prior to the exclusion of light. It is ordinarily accepted, however, that virtually complete adaptation from brightest to darkest can be achieved in rather less than 1 hr. (Piper, 1903). By far the greater part of this change is over in the first 20 min. If the sensitivity of the eye is examined during the process of adaptation, using stimuli of sufficiently low intensity and short duration as not to interfere with the process itself, a curve may be secured which shows sensitivity against time in the dark. Such a curve (Fig. 4) demonstrates, starting from time 0, first a rapid increase in sensitivity, the rate of increase gradually decreasing until a period of some 10 min. has elapsed. At this point the rate of increase accelerates markedly, though it again gradually falls off, becoming almost zero at 1 hr. The change from negligible to rapid rate of change at about 7 min., first noted by Kohlrausch (1922), is ordinarily believed to indicate the point at which cone vision ceases and rod vision begins. This supposition is confirmed by various findings, notably those of Hecht (1921), who, by employing red light as a stimulus, was able virtually to exclude rod vision and thus examine dark adaptation of the cones alone. The early drop in such a curve coincides with that found when the eye is examined as a whole, though the succeeding portion where sensitivity is no longer increasing may be continued *ad infinitum*, the rod-mediated drop being excluded (Fig. 4) (Haig and Wald in Hecht, 1934).

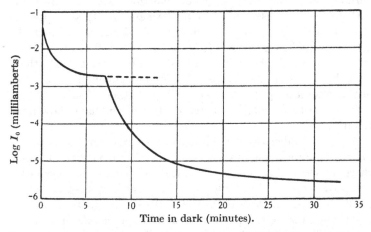

Fig. 4. Dark adaptation curve (after Haig and Wald). The dotted extension of the first curve shows the result of confining the experiment to the cones, by using a centrally fixated, red stimulus. This procedure abolishes the subsequent rod-mediated drop.

Until recently, this simple interpretation of the adaptation curve was generally accepted. Hecht, Haig and Chase (1937) have shown, however, that the curve obtained where the eye is not previously adapted to the maximum brightness possible, but instead to various lower intensities, a procedure which should permit the establishment of curves identical with those for normal dark adaptation, but commencing at various points distant from the extreme low-sensitivity end of the time scale, in fact reveals, at least for moderately bright pre-adaptation illuminations, that the resulting curve shows no such clear inflection into two parts. From this it may be concluded that the transition from rod to cone vision is, under these circumstances, accomplished much more smoothly than where maximum light adaptation is used.[1] It cannot be assumed that the basic Duplicity Theory is seriously contra-indicated.

The study of vision in certain pathological conditions has also shown the essential duality of the retinal process. *Hemeralopia*, more conventionally known as "Night-blindness", is a condition in which vision at low intensities is markedly poorer than in the normal eye. It was early suggested that this state was connected with deficient or absent rod-vision (Parinaud, 1881), and appears to be of two types, one temporary and caused by vitamin A deficiency (*vide, e.g.*, Tansley, 1931), and the other permanent and inherited. It has been clearly shown by Dieter (1929) that night-blind individuals demonstrate an almost complete absence of the second drop in the curve of Fig. 4, their absolute sensitivity suffering a negligible increase after 10 min. have elapsed in the dark.

All the data discussed in this chapter are, more or less, according to their relevance, in favour of the Duplicity Theory; and there is clearly sufficient evidence to justify the generally accepted view of the duality of the retinal process. Certain additional data have been pressed into service in the hope of still further strengthening this virtually impregnable theoretical position. These attempts we shall have occasion to discuss in the next chapter, but we may here remark that such efforts are scarcely necessary, for von Kries' Duplicity Theory was generally accepted before his death in 1929, and most further investigations have only gone to increase this unanimity of opinion.

---

[1] This interpretation appears, at all events, more acceptable than that advanced by the authors in the paper referred to, *q.v.*

CHAPTER 3

The Discrimination of Intensity, and Visual Acuity

The visual discrimination of intensity is a subject which has long engaged the attention of those interested in sensory phenomena. As early as 1760, Bouguer, in Paris, published a book describing his studies in this field, and it is not without interest to note that this early writer arrived at the highly important and entirely correct conclusion that the sensitivity of the eye to differences in intensity is relative, and depends on the absolute intensity at which the differential judgement is made. It is true that neither he nor those who, for the next seventy years, undertook similar investigations, attempted to derive any accurate quantitative conclusions as to the *manner* in which the differential threshold varied with the test intensity. In this they were probably wise, for the apparatus then available for these and like experiments was not such as to permit the attainment of any very high degree of accuracy. The first attempt to establish such a rigid relationship for visual intensity discrimination was that of Fechner (1860), exactly one hundred years after Bouguer. Noting the conclusions of Weber (1835, 1846), and making certain measurements of doubtful accuracy himself, he propounded a hypothesis which, as the Weber-Fechner Law, has persisted almost to the present day. It may be briefly described thus. If I and $I + \Delta I$ are two stimulus intensities such that they produce just perceptibly different sensations, then I and ΔI are so related that the fraction $\Delta I/I$ is constant. Weber had previously (1846) announced the truth of this statement for various senses, though in these he included the eye only to the extent of saying that it held for the visual determination of space, or, more precisely, of length. Fechner was the first to state that it held for visual intensity, however, and here it is on him that the blame must lie, not only for this essentially incorrect "law" itself but also for various quite unacceptable mathematical extensions of it (*vide, e.g.,* Myers, 1928). Numerous investigations since the publication of Fechner's book, starting with those of Helmholtz (1866) and continuing up to the present day, have shown that the equation

$$\Delta I = kI$$

is quite untrue, except for very limited ranges of I, for any sensory mode as yet examined. The classical investigation of the way in which the Weber Fraction, $\Delta I/I$, does vary with I is that of König and Brodhun (1888, 1889), who found that, as I increased from the lowest intensities, $\Delta I/I$ decreased. Their results over the low and medium range of intensities are generally accepted, and indicate that, at the lowest intensities, $\Delta I/I$ approximates to 1, whilst higher up on the intensity scale, and under the best conditions, it falls to as little as 0·006. König and Brodhun also demonstrated that, at the highest intensities, $\Delta I/I$ once more increased slightly from the optimal figure. This rise at high intensities has met with very general acceptance, and has, indeed, greatly influenced most theories of visual intensity discrimination. There is now, however, good reason to doubt its validity, for Guild (1932), Steinhardt (1936) and Smith (1936) have indicated that it appears only when the eye is insufficiently adapted to the test illumination. This has been fully confirmed by Craik (1938). If measurements are made with a test field of adequate size, and sufficient time allowed for the eye to become adequately adapted, $\Delta I/I$ remains optimally small even for intensities so high as to be unpleasant for the subject of the experiment.[1] We may therefore say that, for visual intensity discrimination, $\Delta I/I$ is approximately constant over the range from medium high to very high light intensities, and below these increases very markedly. The discovery that $\Delta I/I$ does not increase at high intensities has led to a complete revolution in theories of brightness discrimination, a point to which we shall return when we come to discuss these. Meanwhile, however, it will be convenient to examine such attempts as have been made to fit the intensity discrimination data into the Duplicity Theory.

[1] At intensities higher still, $\Delta I/I$ may increase somewhat (Steinhardt, 1936). This effect is of no significance, however, as a normal subject is unable adequately to adapt his eye to a stimulus as great as Steinhardt's highest values.

If an attempt is made to measure $\Delta I/I$ for decreasingly small values of I, a point must eventually be reached where I is of such an intensity as to be just below the cone threshold. At this point, it must be supposed, vision previously mediated mainly by cones [1] will give place to vision solely rod mediated. Careful examination of the early data, as plotted, reveals, however, no discontinuity in the curve thus produced which might be held to indicate such a change. Steinhardt (1936), however, has produced some data which, using a field $3°\,44'$ in diameter, in his own opinion and in that of Hecht (1937), do show such a discontinuity. The same investigator also employed a $56'$ field, centrally fixated, and the curves thus produced show no such effect, as would be predicted from the fact that such a field would cover a part of the retina, the centre of the *fovea centralis*, in which cones only would be present. These effects, more particularly the discontinuity in the combined rod and cone curve, are most obvious when a double logarithmic scale is used, that is to say when $\log_{10} \Delta I/I$ is plotted against $\log_{10} I$. It may well be asked whether the earlier data, particularly those of König and Brodhun, do not show a similar effect when similarly plotted. Hecht (1937, Fig. 16) has replotted some of the data for König's eye on similar double logarithmic scales, and maintains that, for those wave-lengths at which both rod and cone functions should be expected there is some evidence for the existence of the required discontinuity. The departure of individual readings from the curves which Hecht draws through them is, however, despite the careful choice of scales, too great to make this sufficiently convincing. Moreover, in order to demonstrate a similar discontinuity in the data of Blanchard (1918), his results have to be plotted on double linear scales, $\Delta I/I$ against I. The significance of this procedure is not altogether clear, for in both cases the two curves drawn by Hecht through the measured data, representing rod and cone functions separately, are the same and are derived from the same equations. The latter are supposed to describe these functions, at least approximately.

At the present time, therefore, it would be safest to say

[1] Owing to their greater number in and near the fovea. This assumes a centrally fixated test patch, as would normally be used.

that the differential threshold data are not such as greatly to strengthen the Duplicity Theory, and it seems probable that, as is not difficult to understand, the change over from rod to cone vision is often gradual, and involves a brief range of intensities at which vision is "mixed"; moreover, it would appear that the slopes of the resulting functions are not always very different.

A considerable amount of work has been done on the eye, using stimuli consisting of regularly alternating periods of light and darkness. If a light source is so arranged that it may be interrupted several times a second, for example, by means of a rotating shutter, it will be found that, as the frequency of interruption is raised, a point is reached at which the light appears no longer to flicker. This frequency is normally known as the critical fusion frequency, or less well as the critical frequency of flicker. It has been argued by Hecht, on the basis of numerous original investigations (for summary, see Hecht, 1937), that the relation between critical fusion frequency and intensity is also such as to indicate the required sudden discontinuity between rod and cone functions. It would appear from other investigations, however, that the point at which this occurs, and the slope of the lines on either side, are dependent upon a number of factors, notably the wave-length of the stimulus. This question is somewhat too involved for, and indeed largely irrelevant to the present discussion, and we may simply note here that other measurements, notably those of Ives (1912), of Allen (1919, 1926) and particularly of Lythgoe and Tansley (1929a), do not confirm Hecht's view, and that the flicker data do not appear to be readily interpretable on the Duplicity Theory. This does not indicate that the latter is wrong, but simply that stable and significant differences between rods and cones do not appear from measurements of this type. We may, therefore, leave these data and turn to those which have been adduced with regard to the acuity of the eye.

It is well known that there is a limit to the capacity of the eye for resolving two points in the visual field separated by an intervening distance whose illumination is to some extent different from that of the points. The resolving power of the eye is normally measured by assessing the angle subtended at the eye by two contours which can just be resolved. Under these conditions, the reciprocal of the angular distance, in minutes, which must separate the two contours, is defined as the *acuity* of the eye. It is a matter of everyday experience that the eye's capacity for such resolution depends not only upon the angle subtended but also upon the ratio of illumination of points to that of the background. The measurement of visual acuity is normally made with test cards, which consist of carefully shaped letters printed in black upon a white background. By subjecting such a card to a varying illumination, and measuring the acuity by noting the size of the letters which can just be read, it is possible to derive the relationship between acuity and the ratio of illumination of background to that of the letters. Where black letters are used it is usually assumed, though not always with justification, that their illumination is zero or nearly so; thus visual acuity is ordinarily measured simply against illumination, that is to say virtually the illumination of the white background. Numerous factors, such as size of background and therefore degree of general retinal illumination, adaptation to the background illumination, and so on, have also to be considered, and profoundly influence the result obtained (Lythgoe, 1932). For our present purposes, however, it might be thought that such factors could be neglected, for it might at least be assumed that all such conditions would be kept constant throughout a given experiment, the illumination only being varied. It would appear that the matter is not as simple as this. The absolute values of acuity realised at any given point are, of course, very largely determined by such factors. There is, however, some evidence that relative values, with which we are here concerned, are also affected. To this point we shall return shortly.

The classical data on the variation of acuity with illumination are those of König (1897). He found, for a normal eye, that if acuity is plotted against illumination, a logarithmic scale being used for the latter in view of the very large range involved and the desirability of giving low values of illumination reasonable prominence, an approximately sigmoid curve is obtained; that is to say, with illumination increasing from the lowest values to the highest, acuity first increased slowly, then, for a considerable range, very much more rapidly, again tailing off at the high illuminations. A point was eventually reached after which further increases yielded no increase in acuity, as is almost always believed.

It would not be unnatural to ascribe the bend at the lower end of the curve to the failure of the cone system of the eye, vision at this point becoming solely rod mediated. Hecht (1934) has, indeed, replotted König's data and drawn through them two curves, one supposedly for rods and one for cones. The justification for this is contained in the data obtained for completely colour-blind individuals, in which cone function is supposed to be absent. From such individuals König himself secured data which might be held to indicate a relation similar to that for rods only in the normal eye. The correspondence is, however, very poor (*vide* Hecht, 1934, Fig. 30). Moreover, the shape of König's curve at both ends of the intensity scale has been seriously questioned by Lythgoe (1932). He has shown conclusively that the tendency for acuity to become constant at the upper end of the intensity scale is due, *inter alia*, to a failure to keep the general retinal illumination approximately similar to 'hat of the test patch. If precautions are taken to ensure approximate equality here, and errors due to the criterion of acuity adopted and to changing diameter of pupil with illumination are also avoided, an almost linear relation continues to the highest measurable intensities. Moreover, though his investigation was not concerned with the eye's acuity over the scotopic range of illuminations, at the extreme low-intensity end of the curve that is to say, such results as he did obtain there indicate similarly that there is no flattening, as was supposed by König. If the inflection in the curve at this end can be avoided simply by controlling the state of adaptation of the rest of the retina, and taking certain other quite elementary precautions, it is clearly impossible to ascribe it to a changeover from cones to rods, as Hecht has done right up to the present time (1934, 1937). In the later of these papers Hecht emphasises the discontinuity in König's curves by replotting the results on a double logarithmic scale, and making certain other adjustments.

If the flattening out is not real, however, this procedure is not convincing.

There are, moreover, very important implications of Lythgoe's finding for the theory of visual intensity discrimination and acuity, to a discussion of which we may now turn. It will be convenient to consider the latter first.

It is clear that any theory of visual acuity based strictly upon the laws of geometrical optics is quite inadequate to explain the data which we have just discussed. Thus an hypothesis envisaging simply an inverted and reduced image of the external object field falling upon a retina composed of discrete photosensitive elements, each with its discrete neural connection with the higher centres, cannot attempt to account for the simple fact that acuity is intimately connected with illumination. It is, in any case, generally supposed that the lens system of the eye suffers from certain optical defects, and because of these and of the inevitable diffraction the retinal image is at best a blurred and distorted counterpart of the object producing it. Various attempts have been made to account for the variation of acuity with illumination, taking this into account. Thus, rejecting the highly over-simplified view that two points in the visual field will be resolved when their images fall upon two cones separated by one unstimulated one, Hofmann (1920) has assumed only that the centre one will be stimulated somewhat less intensely than the outer ones; this owing to the retinal image not being adequately sharp. He points out that, this being the case, the effect of raising the illumination of the points will only be to raise the general illumination of the relevant area of the retina. The *relative* distribution will thus remain the same, though the absolute level will rise. Now, over the middle and high range of intensities at least, the Weber-Fechner Fraction ($\Delta I/I$) is at least approximately constant, and thus the differential intensity sensitivity of the eye will not increase sufficiently to account for the marked change in acuity over this range. At the lowest intensities, where $\Delta I/I$ is changing very rapidly indeed, this simple theory has much to recommend it, but for the higher intensities it demands some amplification. In order to do this, Hofmann extends Hering's (1907) hypothesis of the reciprocal action of the retina, advanced to account for the phenomena of Simultaneous and Successive Contrast, to cover the acuity data additionally. In simple terms, the increase in acuity as illumination is augmented is held to be due to the subjective difference between the most brightly illuminated and the least illuminated parts of the retina being increased by Simultaneous Contrast, or, as Hering calls it, Spatial Induction. Wilcox (1932) has extended this theory, and subjected it to certain experimental tests. In these, two rectangular objects, whose illumination was dissimilar from that of the background, are approximated until the apparent distance between them is equal to their apparent width. Comparing these measurements with the actual dimensions of the objects, it is possible to calculate the amount of irradiation at various illuminations. Wilcox shows quite conclusively that this relation is such that Simultaneous Contrast might well account, at least to some extent, for the acuity data; though no explanation of the variation of irradiation with illumination is offered.

Theories of visual acuity based on the fact that, in some animals, there is a migration of epithelial pigment as the illumination of the retina is increased have been advanced, notably by Broca (1901). This phenomenon, if it occurs at all in man, does so only to a very small extent, however (Arey, 1915 a, b), and this hypothesis is not ordinarily held in much favour. It is of interest to note, however, that Crawford (1937) has recently produced some evidence arguing the existence of the effect, and it appears not inconceivable that this phenomenon may exercise a secondary influence on visual acuity.

According to Hecht's theory of visual acuity (1928 a), different rods and cones are conceived as having widely different absolute thresholds; the differences in threshold being distributed, throughout a given area of the retina, in a manner similar to the distribution of some characteristics in other populations. Thus, if the number of elements with a given threshold is plotted against retinal illumination, what is roughly a Gaussian distribution curve is obtained, demonstrating that a majority of the photosensitive elements have thresholds approximately midway between the highest and the lowest. Such a distribution is thought to hold both for rods and for cones, but in addition it is legitimately assumed that the rods as a whole have lower thresholds than the cones.

If this were the case it is clear that, at the lower illuminations acuity would be poor owing to the small number of elements which it would be possible to stimulate, and, on the average, the great distances between them. As illumination is increased, the *active* population of a given area would steadily approximate to the *potential* or full population. At a certain point, however, a given fractional increase in illumination would no longer increase the active population by as great a fraction as it would have done had it occurred lower on the scale, as at this point the distribution curve is once more commencing to descend. Thereafter acuity should become steadily more constant. Hecht derives his distribution curves from the acuity data, those of Roelofs and Zeeman (1919) and of König (1897) being used in the paper referred to (Hecht, 1928 a). As the acuity v. illumination curve ordinarily shows a much more obvious inflection at the lower than at the higher intensity end, the distribution curve is somewhat skew, as would be expected. In order to discover how many retinal elements are functional at any given illumination, rather than the number having their thresholds at that illumination, curves are plotted showing the integral form of the distribution curves discussed. Two further curves, for rods and cones respectively, are thus obtained. Hecht now observes that these latter curves cover the acuity data well. As the former curves must essentially have been derived from the acuity data by the reverse process to that described, this is not at all surprising. It is interesting, therefore, that Hecht, on producing the integral curves, notes that they are described by the equation

$$KI = x^2/(a-x), \qquad \ldots\ldots(1)$$

where K and a have different values for the separate rod and cone curves. A little later in the development of this theory, the visual acuity data of König (1897) are plotted, with three curves drawn through them. All three curves are described by the equation

$$KI = x^n/(a-x), \qquad \ldots\ldots(2)$$

and are for $n = 1$, 2 and 3 respectively. Hecht notes with apparent surprise, however, that the best fitting curve is that where $n = 2$. It is not necessary to point out that this is equation (1), and that this procedure indicates no more than that the necessary mathematical manipulation has been correctly carried out.

It is clear from this that Hecht's mathematical treatment of the acuity data need not here demand further consideration, and that the precise form of his derived equations is most probably not of great consequence. We may, however, attempt to interpret Hecht's theory with more latitude, and to see what consequences follow from the simple notion that the retinal photosensitive elements are of different sensitivities, and that these are distributed in the manner discussed. One point emerges instantly. If any reasonable distribution curve whatever is assumed, a point must be reached at which a given fractional increase in illumination will bring about a smaller increase in number of photosensitive elements than if the same fractional increase had been made at a lower absolute illumination. In short, on this theory, if either $\Delta I/I$ or acuity is plotted against illumination, there must be an inflection at the upper intensity end. As we have just seen, however, there is good reason to suppose that, in both these curves, the bends can be greatly reduced or, for $\Delta I/I$, completely avoided, by taking precautions with regard to the adaptation of the eye, etc. At the lower intensity end a similar bend must, of course, be predicted, but the experimental position is confused by the fact that here photopic vision is being replaced by scotopic before such a point can be reached.

It seems probable, therefore, that Hecht's first theory may be abandoned on these data alone. There are, in any case however, two further consequences which render its acceptance difficult. First, visual acuity must be, on such a theory, limited by what in photographic terms may be called the "grain" of the retina. As Lythgoe (1932) has pointed out, Hecht's theory inevitably assumes far too high a cone population at the fovea, for over the major range of illumination many of the cones must be lying idle. As we shall shortly see, visual discrimination is far superior to anything which might be predicted from a consideration of the grain of the photosensitive surface, that is to say, for the eye, the number of cones per unit area. Thus no theory can be accepted which does not take into account this extraordinary discrimination, still less one which assumes a grain actually inferior to the best possible at all but unrealisably high illuminations.

One further criticism may be made. Both this theory and that of Houstoun (1932) take no cognisance of what is now recognised to be one of the most fundamental properties of all nerve endings, namely, that the frequency of nerve impulses varies with the intensity with which they are stimulated (e.g. Adrian, 1928). There is no doubt whatever that this is true for the eye, as we shall see below, as it is for every other sense organ investigated.[1] To some extent Hecht has realised this, and in a later exposition of visual theory allows the frequency of nerve impulses a minor place (Hecht, 1934).

The theory of visual acuity due to Hartridge (1923) is unique in that it attempts to account for the fact, already mentioned, that visual acuity under certain circumstances can so far outstrip the grain of the photosensitive surface. Hartridge points out that the size of the foveal cones is, if an average of all values given by different observers is taken, such that their diameter is of the order of $3 \cdot 2\,\mu$. Supposing the focal length of the eye to be 15 mm., then a cone diameter of $3 \cdot 2\,\mu$. yields an acuity corresponding to an angle of 44 sec. of arc subtended at the eye; this assuming that, for resolution to take place, two cones must be stimulated with one unstimulated (or less stimulated) between them; or, of course, vice versa. Even if the diameter of the cones were as low as $2\,\mu$., the angle separating the centres of two illuminated objects should not be less than 28 sec. of arc, for resolution to take place. The acuity of the eye is, however, often much greater than this. For example, Baker and Bryan (1912) have shown that the accuracy with which two lines can be set into coincidence (vernier acuity) is between 8 and 10 sec. of arc; whilst other observers have reported even smaller figures (for summary, see Hartridge, loc. cit. p. 55). It is quite clear, therefore, that no simple geometrical theory, even if extended to cover

---

[1] Except possibly the ear: vide infra.

the acuity and illumination data, will account for these results. Hartridge's theory does so by assuming that the so-called optical imperfections of the retinal image are actually turned to advantage. Proceeding on the assumption that the eye suffers from various forms of chromatic aberration in the same way as do other optical instruments of comparable aperture, as we have already seen (Chapter 1) to be almost certainly the case, Hartridge calculates the form of the retinal image produced by a bright narrow line on a dark background. If the intensity of light falling upon a cone corresponding to the centre of the image of such a line is equal to 100, then the two cones immediately on either side are subjected to an intensity of 31 p.c.; those on either side of these to 9 p.c. and so on to the sixth row of cones which are illuminated to the extent of 1 p.c. In short, a much larger number of cones is stimulated by such a source than would be thought to be the case according to the rules of geometrical optics. If it is now assumed that the line is broken, and the ends not entirely coincident, as in the vernier acuity experiment, then the distribution of light on the various stimulated cones will be greatly changed; this even if the departure from coincidence is so small as to correspond to a shift of a given part of the image of considerably less than the diameter of one cone. Assuming this state of affairs, then, the eye's acuity is limited only by its differential *intensity* sensitivity. Hartridge's calculations yield values for this, for the small retinal area involved, which seem very likely. Moreover, his experiments reveal, as would be expected on this assumption, that if the intensity of the test object is constant acuity is better than if it is varying. It is clear that only a very small extension of this theory is necessary to account for the variation of acuity with illumination, for it is known that differential sensitivity varies with it, and thus acuity should do so in roughly the same manner. Within quite reasonably close limits, as we have seen, this is true. Lythgoe (1932) makes the criticism that, whereas the brightness discrimination of the eye improves only up to a certain point, say 3 f.c., and then deteriorates, acuity continues to improve much further than this. As we have seen, the inflection in the upper end of the $\Delta I/I$ curve is largely a matter of adaptation, and is not present where adequate experimental precautions are taken. It cannot, however, be denied that intensity discrimination does tend to become constant at a relatively low intensity, and, whilst a small improvement is usually found thereafter, visual acuity, on the same scale, continues to improve very markedly. No doubt, therefore, Hofmann's (1920) suggestion, already discussed, that the effects of Simultaneous Contrast also increase with illumination should also be included, and may well account for this continued improvement. It will at once be seen, moreover, that Hofmann's hypothesis is in complete harmony with Hartridge's formulation. At the present time a theoretical picture of visual acuity combining these two concordant hypotheses appears the most attractive. It is, of course, not impossible that rods and cones vary *inter se* in threshold, as Hecht suggests; but such variations as exist are probably extremely small, perhaps negligible, in comparison with those which he postulates. If this is so, it is unlikely that these minor individual differences are important for the theory of visual acuity.

It is sufficiently clear from the foregoing discussion that visual acuity and visual intensity discrimination are intimately connected. We must shortly come to consider theories of the latter, and not until then will our picture of acuity be complete. This examination we have left to the end of the chapter in order to stress the intimate relation between the two effects. Before it can be undertaken, however, consideration must first be given to certain data which have accumulated to show the relation between the intensity of stimulation of the eye and the nature of the discharge in the optic nerve.

It has been known for many years that a message travelling in peripheral nerve will reveal its presence by certain electrical changes, ordinarily of a fairly small order (*vide, e.g.*, Adrian, 1932). Now, if the nerve attached to a muscle is stimulated, for example by pinching it, the muscle will contract. If two fine wires are placed on the nerve and these connected to some electrical recording instrument of adequate sensitivity and speed of response, such as a string galvanometer, it will be seen that whenever a message passes to the muscle, causing it to contract, the instrument reveals a series of transient electrical potentials appearing across the electrodes on the nerve. The type of nerve trunk used in this simple experiment consists of a very large number of fibres, and a record of the potentials thus obtained, against time, would present a very erratic picture; this because the messages in individual fibres will not necessarily arrive at the electrodes at the same time. The overall result will be, therefore, that the messages in different fibres will sometimes summate whilst at other times they will not. Much of the knowledge which has accumulated about the nature of the messages in both sensory and motor nerves, there being from this point of view little difference between the two, has arisen from the examination of messages passing in one fibre at a time. Sometimes this is accomplished by cutting down a nerve until only one unsevered fibre remains (Adrian and Bronk, 1928), and sometimes by the employment of a receptor attached only to one fibre (Matthews, 1931). Under these conditions, the electrical picture is very much simpler.

On stimulating a single fibre with electrodes placed on it in the manner described, and obtaining a record by means of a moving film camera and a suitable recording instrument, one great simplification is at once apparent. The electrical indication of the message now consists of a series of diphasic[1] potential changes, all of almost exactly the same shape and size. With certain exceptions with which we are not here concerned, this invariability in both size and form is complete. No matter what stimulus is applied, nor what its strength, neither is in any way changed.

From these and much other data, which it would be quite out of place to discuss here (see, however, Adrian, 1928, 1932), it has been concluded that the messages in nerve fibres, both sensory and motor, consist of a series of individual units, to which the name *impulse* is ordinarily given. Each impulse produces a single transient variation of electrical potential, complete in itself, to which the term *action potential* is usually applied.

[1] That is to say, such that the voltage to the recording instrument, initially zero, first becomes (say) negative, then positive, and then returns to zero.

From the study of these action potentials a great deal of information about the sensory mechanism has been gained. It is almost invariably believed that each impulse produces an action potential and that there are no messages in nerves other than those composed of these impulses. Weiss (*e.g.* 1931) has, it is true, produced an hypothesis to the contrary, but at the present time this is ordinarily accepted as being the case (Adrian, 1932, pp. 9–21). In the ensuing discussion we shall certainly take it for granted.

The invariability in size and form of the action potential will immediately provoke an interesting question. It may well be asked in what dimension the nervous message does vary, in response to a change in intensity of the stimulus. Provided, as always, that the investigation is confined to the impulses in a single fibre, this question can now be answered, at least for all sensory and motor nerves thus far investigated. It appears to be the case that, provided certain limits are not exceeded, an increase in the intensity of stimulation, either of a sense organ or of the nerve directly, results in an increase in the *frequency* of the nervous impulses, an increase, that is to say, in the number of individual impulses passing in unit time. Where a number of nerve fibres are concerned, a further result of increasing the stimulus intensity from a low value to a high is to cause impulses to pass in fibres which, at the low intensity, have been inactive. With the possible exception of the auditory nerve, which we shall consider later, the two dimensions of the message corresponding to the intensity dimension of the stimulus are, then, frequency and number of fibres active. With this in mind, we may turn to an examination of the data deriving from the study of the impulses in the optic nerve.

The first reported investigation of this type is that of Adrian and Matthews (1927 *a*, *b*, 1928). The animal used was the conger eel, which possesses certain advantages from the anatomical point of view in that the optic nerve is unusually long and accessible. It was not, however, possible to cut this nerve down to a single fibre, nor would such a procedure have been of much value, for the vertebrate eye possesses, in the retinal synapse layer, so complicated a system of interconnection of fibres that a single fibre here would certainly not reflect the activity of a single photoreceptor, nor indeed even that of a reasonably small number. The single-fibre technique has, however, been successfully applied to the eye of the horseshoe crab, *Limulus polyphemus*, by Hartline and Graham (1932), and several investigators (*e.g.* Demoll, 1914) have shown that this eye lacks internuncial neurons and has no structure at all comparable to the

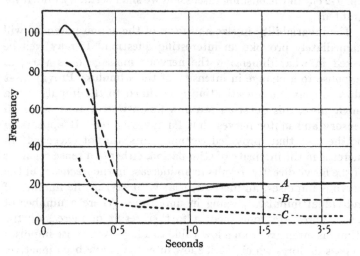

Fig. 5. The relation between frequency of impulses and time after onset of stimulus, for a single fibre from the eye of *Limulus* (after Hartline and Graham).

      A = for stimulus intensity of 63,000 metre-candles,
      B =   ,,   ,,   ,,   630   ,,   ,,
      C =   ,,   ,,   ,,   63   ,,   ,,

vertebrate retinal synapse layer. The nervous discharge here, therefore, should be of the greatest simplicity and the clearest significance.

Using such a preparation, the action-potential record obtained (Hartline and Graham, 1932) may be briefly described as follows. On applying the light stimulus to the eye, a typically single-fibre discharge commences after a brief latent period of the order of 0·1–0·2 sec. Initially, the discharge is of high frequency, but rapidly sinks to a lower level, which is maintained quite regularly until the light stimulus is removed. When this occurs, the discharge persists for perhaps 0·1 sec., and then ceases. The effect of changes in stimulating illumination on such a response is of the greatest interest. A higher stimulus intensity results in both the initial maximum frequency and the final steady value being increased, in a manner which has been found to be typical of

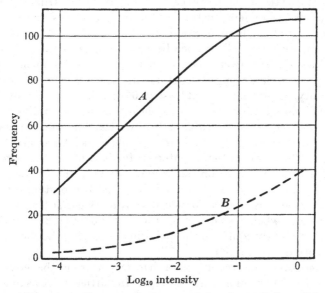

Fig. 6. The relation between Frequency of impulses and Log. light intensity, for a single fibre from the eye of *Limulus* (after Hartline and Graham). Curve *A* = Frequency of initial maximal discharge. Curve *B* = Frequency of discharge 3·5 sec. after the onset of the stimulus.

other sense organs (*e.g.* Adrian and Zotterman, 1926 (frog muscle receptors); Bronk, 1929 (frog skin and muscle receptors); Matthews, 1931 (mammalian muscle receptors)). At the lower intensities the discharge frequency is lower, and in addition the latent period at the beginning of the discharge is somewhat lengthened. At the lowest stimulus intensities the later discharge also tends to become somewhat irregular. These data are clearly seen from Figs. 5 and 6, taken from the paper referred to.

Before we attempt a theoretical interpretation of these results
it is as well to consider how far we are justified taking as general,
more particularly for the human eye in which we are mainly
interested here, results obtained from a specific eye, moreover
that of a primitive organism. This question cannot, at the moment, be finally settled. The close correspondence between these
results and those established for other mammalian end-organs
strengthens the belief, however, that we have here a true, though
no doubt incomplete, picture of the earliest stages in the development of the human visual sensation.

From Fig. 6 it will at once be seen that, for a range of intensities of rather over 1000 : 1, the frequency of the initial discharge is proportional to the logarithm of the stimulating
intensity. Over a range of at least 10,000 : 1 the frequency of the
later discharge varies with the intensity, though according to a
somewhat more complicated relation. Hartline and Graham
state that, in some experiments this range may be as great as
10^6 : 1. This is of particular interest, for it is at once apparent
that, if a single receptor can respond at different frequencies
over so enormous a range, there is little need to postulate that
receptors are distributed for sensitivity, as Hecht still states to
be the case when accounting for the acuity data (1937, p. 277).
We shall shortly return to this point in considering the theory of
intensity discrimination.

Much other interesting data has arisen from the electrophysiological examination of the eye. In particular, Adrian and
Matthews (1928) have shown quite conclusively that the latent
period of the discharge in the optic nerve of the vertebrate
conger eel is very largely influenced by the interaction of retinal
neurones, which, as we have seen, must take place in all but the
simplest eyes, certainly in all mammalian eyes. The precise
significance of this for the theory of vision is, perhaps, not yet
clear; but we may in passing note that attempts to deduce
mathematical relationships governing the reactions of retinal
photoproducts are, if based on data derived from the immensely
complicated human eye, almost certainly doomed to failure. This
is the more evident when it is remembered that, as we have seen,
the shape of both the intensity discrimination and the acuity

data curves for the human eye is profoundly influenced by the level of adaptation of the retina *as a whole*; a state of affairs easily understandable in view of Adrian and Matthews' demonstration, for example, that the length of the latent period of the neural response is dependent upon the size of the retinal area stimulated. There is important evidence in this work that the locus of this interaction is the retinal synapse layer. The presence of this layer will render the electrophysiological study of the mammalian retinal process extremely difficult.

With these facts in mind we may now complete our theoretical picture of the eye as a light-sensitive instrument by considering the theory of intensity discrimination.

It will be clear from the data which we have already discussed that, in all probability, the frequency of impulses in the optic nerve fibres will constitute one important clue in the discrimination of intensity. This matter is no longer in dispute, and it may be safely-said that at the present day this factor must certainly be included in any theoretical interpretation. Historically, however, it is of some interest to note that theories of vision have been proposed based solely on the supposition that the photoreceptors differ in threshold, but that otherwise their operation is "all or none". For the eye, this supposition was first made by Lasareff (1923). It was later treated quantitatively by Hecht (1928 *a*), and in a somewhat different manner by Houstoun (1932 *a, b*).

We have already seen that, in order to account for the acuity data, Hecht has postulated that the receptors of the eye are distributed for threshold of response, and that their thresholds yield a somewhat asymmetrical distribution curve of fairly conventional form. Now such a distribution can also be made the basis of an intensity discrimination theory, for if an increase in stimulus intensity yields an increased number of active receptors, an assumption inseparable from the supposed distribution of thresholds, then it may be that such discrimination takes place on the basis of the total active receptor population. This is the theory propounded by Hecht (1928 *a*) in a paper designed primarily to formulate a theory of acuity, but whose implications for the theory of intensity have also been stressed (Hecht,

1934). In this latter description the role of nerve-impulse frequency is also considered; such a compound theory as this demands some examination.

It will at once be seen that the sensitivity distribution basis of intensity discrimination makes precisely the same demands on the shape of the curve connecting illumination and differential sensitivity as it does on the acuity curve. That is to say, the curve must show some inflection at the high intensity end if any reasonable distribution curve whatever is supposed. Recently the evidence has been rapidly accumulating that, with suitable conditions of pre-adaptation of the eye before the high intensity measurements are made, this is not the case. This was first shown for bees, and later for the fruit-fly, *Drosophila*, but we are here more interested in human data. We have already seen that the existence of this bend has been seriously questioned, and Craik (1938) has shown almost certainly that the bend can be avoided by suitable adaptation. The immediate theoretical consequence of this is either that the "distribution of thresholds" hypothesis must be abandoned altogether or that it is necessary to assume a distribution curve such that the descending high-intensity portion falls at illumination values so high as to be almost unrealisable, and which would quite certainly be deleterious to the eye if any attempt were made to reach them. In brief, a large number of the cones are so insensitive that no actual stimulus is adequate to excite them. The absurdity of this supposition is at once apparent.

It would appear from the later formulation of the intensity position which Hecht (1935) has published that this theory has been abandoned. This later paper, however, is concerned mainly with a mathematical formulation of the discrimination data, and from it it is difficult to see whether the distribution theory has been abandoned or merely modified. It may well be that the latter is the case for, in a still later summary (Hecht, 1937), the author appears to maintain that the receptor distribution hypothesis may still be considered. Thus, referring to the distribution of thresholds curve, Hecht states (*ibid.* p. 277):

"This statistical treatment of visual acuity remains to-day much as it was proposed."

Several difficulties at once arise. If, in order to account for the acuity v. illumination data, it is essential to assume that the receptor thresholds *are* distributed in this way, it is quite impossible wholly to neglect this supposition in considering the intensity discrimination material. In short, if the eye's acuity is governed by the number of operative receptors, then so, at least in part, must its intensity discrimination be. We have already seen that there is good evidence for abandoning this concept in discussing acuity; in the effect of adaptation in straightening the intensity discrimination curve we have equally good evidence for relinquishing it here.

We have earlier noted that a modification of this theory has been proposed by Houstoun (1932 a, b). On Hecht's formulation discrimination is mediated by the total number of elements active. On Houstoun's version, the supposition is made that it is governed by the rate of entrance of active elements. Precisely the same strictures as apply to Hecht's form apply here also, however, and we need go no further into these possibilities.

At the present time, therefore, we are compelled to assume that intensity discrimination takes place on the basis of the frequency of optic nerve impulses, with receptor threshold differences mediating at most some minor part of this function at very low stimulus intensities; for it may be taken that all thresholds, for cones and rods independently, are approximately similar, and therefore that only at threshold light values will such differences as exist appear of importance. It may further be supposed that, over the relevant range of intensities, the change over from rod to cone vision may offer some further clue; though as has been said earlier this is probably more gradual than has sometimes been thought.

With one further point it is desirable to deal. In considering intensity discrimination it is possible to make a number of assumptions with regard to the nature of the underlying photochemical mechanism. Thus in the retinal photoreceptor system we may suppose the existence of (a) an inactive light-sensitive substance, which is changed by the absorption of light into an active substance which is responsible for the initiation of the events leading up to the nervous impulse, and (b) an arrangement which maintains the supply of the inactive substance, since otherwise the whole process would eventually cease. This simple reversible photochemical system has been made the basis of all Hecht's theoretical treatment of photosensory data (*e.g.* Hecht, 1934). Now it may be assumed that there exists, for every stimulus intensity, high or low, to which the eye is fully adapted, a certain photochemical equilibrium, such that the concentrations of sensitive material and decomposition products remain constant. This is an essentially stationary state. We may further suppose that the difference between two dissimilar equilibria may be stated as consisting of a given amount of photoproduct, that is to say, of the substance produced in the receptor cells by the action of light.

Now, when the intensity discrimination of the eye is being measured, it is clearly necessary eventually to arrange two stimuli in such a way that they are just discriminably different. This may be done in such a way that the eye views a bipartite test field, one-half of which is adjusted independently of the other until the required condition of just perceptible difference is achieved. In this case the resulting measurement is known as the *simultaneous* differential threshold. Alternatively, the two intensities may be so arranged that they alternate in time, permitting the assessment of the *successive* threshold. In the latter case the two photochemical equilibria to which the two intensities I and $I + \Delta I$ correspond will also alternate. We are thus left with two possibilities. We may assume that the discrimination taking place is due either (*a*) to the difference between the two equilibria involved, the difference, that is to say, between the two steady states, or (*b*) to the change from one equilibrium to the other. Each of these assumptions has been made, at different times, by Hecht, as the basis of a theory of intensity discrimination. Thus, according to his earlier theory (1924, 1928) the eye will just discriminate between two intensities when their corresponding stationary states differ by a constant amount of photoproducts. This formulation predicts a rise of $\Delta I/I$ at high intensities, and has had therefore to be abandoned (Hecht, 1935) in favour of the second assumption, namely, that it is the change between the two which is of im-

portance. More precisely, in Hecht's second theory the following assumption is made: that, for the intensity $I + \Delta I$ to be just
distinguished from the intensity I, the initial velocity of the
reaction of decomposition of photosensitive material, on the
addition of the extra intensity ΔI, must be constant irrespective
of the magnitude of I. The curves yielded by the resulting
equation appear to fit the intensity discrimination data of
Drosophila, of the bee, and of the clam *Mya*, the experimental
material in the last two cases being that of Hecht himself. For
all these organisms, the procedure is such that the successive
threshold is measured. Now, at first sight, it would appear that
the measurement of the simultaneous threshold should produce
quite different results. Moreover, it is this type of measurement
which has usually been made for the human eye. Hecht has
attempted to fit the human data into the equations derived. He
finds, as would be expected, that they do not fit those where
$\Delta I/I$ tends to increase for high values of I, and as we have said
later investigations have shown this rise to be spurious. But
normally, the fit is good, despite the apparently incorrect
experimental procedure. To account for this Hecht points out
that the procedure employed in simultaneous threshold experiments is ordinarily such that only one side of the bipartite field is
varied in intensity. Thus, starting from a point where both sides
are the same, the intensity of one is varied until the two are just
discriminably different. When these conditions are observed, the
change between the two stationary states may well be the important factor. Where continued examination of the differential
field reveals that the perceived brightness difference is maintained, a condition difficult to fit into this theoretical formulation, Hecht assumes that eye movements take place, as Helmholtz originally pointed out (Helmholtz, 1924, 2, pp. 179, 237,
266 and 281). Thus he supposes that fresh parts of the retina are
exposed to the higher intensity, though the general state of
adaptation of the retina remains at a level corresponding to the
lower intensity. It is clear that, for these conditions to be fulfilled, highly specific experimental arrangements must be employed. Moreover, it is difficult to see why, if an equally divided
bipartite field is used, as would ordinarily be the case, the general

adaptation level of the retina should correspond to the lower of the two illuminations rather than to the higher, or to a state midway in between.

A simple extension of Hecht's theory would appear to cover this condition. It requires to be assumed, as in his case, that some eye movements take place, but it is relevant to consider the state of adaptation of the retina at an imaginary moment in time before these have occurred. If, as we have already decided, the test field is divided into two equal or unequal parts whose intensities are I and $I + \Delta I$ respectively, it is extremely likely that those parts of the retina exposed to the illumination I will be very slightly less bright adapted than those exposed to the higher illumination $I + \Delta I$. The *overall* state of adaptation of the retina is thus assumed to correspond to a point midway between I and $I + \Delta I$, but those parts specifically illuminated by the latter are adapted to a slightly higher intensity than are those illuminated by the former, and lower of the two brightnesses. It is not assumed that the state of adaptation achieved in either case will be equivalent to the final steady state reached by continuous adaptation to either intensity, but simply that a small difference in state of adaptation will none the less be present. General considerations make this supposition not unlikely.

Immediately eye movements have taken place, some parts of the retina previously exposed to I will be illuminated by $I + \Delta I$, or *vice versa*. Under certain test conditions (for example, where one test field is within the other), both these changes of illumination will operate. Where an equally divided, bi-partite field is used, one or the other will occur, depending on the direction of eye movement relative to the two half-fields. Irrespective of the test conditions, however, any movement must cause in some retinal area a reaction towards a new equilibrium, either higher or lower, to be initiated. It is probable that the tendency towards a state of adaptation equivalent to a higher illumination will be the more important of the two, though this is not certain as yet. This movement towards the new adaptation state, regardless of whether or not it is in fact finally achieved, is, we may suppose, the clue for the intensity discrimination judgement, just as it was for the successive threshold case.

This formulation appears to have the twofold advantage of placing simultaneous discrimination on the same basis as successive, and of doing so without the necessity of assuming highly specific experimental arrangements as must be the case for Hecht's hypothesis. The virtual independence of the simultaneous differential threshold of minor modifications of test technique implies the necessity for a theory fulfilling these conditions.

Continued experimental activity will no doubt assist in illuminating the obscurity here. At the present moment we may say that, interpreted broadly, Hecht's suggestion that the recognition of intensity differences may "involve initial rather than final effects" (1935, p. 779) is one well worthy of consideration; it may well prove to be a clarifying concept in this difficult field of sensory physiology.

CHAPTER 4

Colour Vision

In the preceding chapters we have discussed such theoretical mechanisms as have been proposed to account for the eye's discrimination of intensity and of space. With few exceptions, the stimuli employed in the determination of the eye's characteristics in these respects have consisted of white light, at various intensities; that is to say, of radiant energy of many wavelengths, simultaneously operative, the magnitude or spatial characteristics of the stimulus only being altered. In general, it has been customary to employ for these experiments either sunlight itself or one of the various forms of artificial sunlight, whose spectral distribution curves are reasonably closely known. In the present chapter, the reaction of the eye to stimuli, the greater part of whose energy lies in a small range of wave-lengths, will be discussed. Such a stimulus, provided that it lies within certain limits of wave-length, is characterised to the normal eye by the possession of a definite hue. Though it is customary to discuss such matters as these under the general heading of "colour vision", the term colour should, in any satisfactory terminology, include the white-black continuum with which we have been primarily concerned up to the present. The term *achromatic* colour should be used to differentiate this series from that produced by the spectrum; for the components of the latter the term *chromatic* colour should preferably be reserved (Troland, 1922, 1934).

The dispersion of the components of white light, either by a prism (constituting the so-called prismatic spectrum) or by a diffraction grating (producing the diffraction spectrum) reveals to a majority of observers a highly significant difference between the chromatic and the achromatic colours. I have earlier used the term "continuum" in describing the latter, for the passage from one end of this series to the opposite involves that one traverses an essentially continuous series of greys, each being different from that last seen only by a "just noticeable difference". The spectral series is to be differentiated from this, for the passage from one end to the other shows that the mode of variation changes sharply at certain reasonably well-distinguished points.

The solar spectrum, provided that it is sufficiently bright, is visible from approximately 400 to 780 mμ. (Troland, 1934), though there is some dispute as to its precise limits. Proceeding from the shorter wave-lengths to the longer, the following series of spectral hues may be differentiated: violet, blue, blue-green, green, yellow-green, yellow, orange, and orange-red. Three psychological primaries may be distinguished here, namely, yellow at 575·5 mμ., green at 505·5 mμ. and blue at 478·5 mμ. (Westphal, 1909). It is evident at once that the psychological primary red finds no place here, the spectral reds being more yellow than this. Such a hue can only be obtained from the spectrum as a heterogeneous stimulus, the two components being selected from the spectral extremities. Other choices from the spectral extremes yield a series of extra-spectral purples, which also find no place in the simple spectrum (e.g. Smith, 1925).

It is well known that certain spectral colours are complementary to others. By this it is meant that if two stimuli are allowed to operate simultaneously upon the eye, adjustment of their intensities separately will produce a grey or white stimulus no longer characterised by the possession of hue. The relative brightnesses of the two components complementary to one another throughout the spectral scale have been determined by Sinden (1923). From his figures it may clearly be seen that the proportions of the two components are closely dependent upon their wave-length. Thus, spectral orange at 609 mμ. is complementary to the blue-green at 493·5 mμ., grey resulting when their luminosities are in the ratio 39·3 (orange) to 35·7 (blue-green); or very approximately 1 : 1. For yellow at 570·5 mμ. and extreme violet to yield grey, their ratios must be approximately 40 : 1. Intermediately, intermediate ratios obtain. This curious fact is of great significance to the theory of colour vision. It is at once evident to any observer that the various hues of the

spectrum are not equally saturated.[1] Thus the mid-spectral
yellows are obviously much less saturated than are the reds and
violets of the spectral extremities. The reason for the dispro-
portionate brightnesses of yellow and violet required to yield a
complementary grey was thought by Helmholtz (1924) to be
connected with this fact. Thus he supposed that with a violet
stimulus most of the sensation resulting was "violet", whilst the
unsaturated yellow produced a large fraction of essentially
"white" sensation. For a complementary mixture, therefore,
enough yellow must be employed for the fractional "pure hue"
to balance the high percentage of "pure hue" of the violet. This
notion requires that it be assumed that a pure yellow hue, if it
could be obtained, would be as saturated as any other pure hue.
Such considerations have led to the production of "colorimetric
purity curves", notably by Priest and Brickwedde (1926). The
procedure employed by these investigators may be briefly
described. A given intensity of white light is permitted to
illuminate an area in the visual field. To one-half of this, homogeneous spectral light is added, the intensity of the white light
being simultaneously reduced to ensure brightness compensation.
In this way, it is possible to determine the minimum amount of
spectral light required just to colour one-half of the field. Such
a procedure necessitates the assumption that the actual amount
of *fully* saturated colour which has to be added to reach threshold
value is the same for all wave-lengths, for only so can the saturation of the hue added be taken to be inversely proportional to the
amount which it is necessary to add. For these assumptions
there is no valid evidence. Thus the supposition that equal
quantities of all saturated hues are required to make a threshold
change in the colour of a previously white field may well be unjustifiable. In so far as one is able to make such a judgement, it
would appear likely from introspective evidence that yellow,
even if it could be secured in full saturation, would yield a
sensation approximating more closely to white than, for example,

---

[1] For purposes of definition, the *saturation* of a stimulus may be regarded
as inversely proportional to the amount of white light mixed with the given
pure hue. Thus pink is an unsaturated red. In lay terminology, unsaturated
colours are said to be "pale".

that produced by stimuli from the spectral extremes. To these
important considerations we shall return later, however. For the
present, it may be noted that the homogeneous stimuli of the
normal spectrum are by no means equally saturated.

In addition to variations of saturation, the spectrum of white
light reveals certain differences of brightness. It is evident to
all observers that, on passing from one spectral extreme to the
opposite, the brightness of the resultant sensation steadily
increases until, for a bright spectrum, a wave-length of some
554 mμ.[1] is reached, after which the brightness once more falls,
reaching a minimal value at the opposite spectral extremity
(Gibson and Tyndall, 1923). To these matters we have already
given some consideration in Chapter 2, when considering the
evidence for the Duplicity Theory of the retinal process. It was
there noted that the brightness distribution of the spectrum was
not independent of its overall illumination, the maximum being
shifted to 511 mμ. at extremely low intensities, that is to say for
scotopic vision. Under these conditions, the spectrum is seen as a
colourless band of light. On the Duplicity Theory this is held to
be because vision is then solely rod mediated. We are here concerned, however, with colour vision, on this theory a cone-
mediated affair; we shall, therefore, confine our attention to the
brightness relations of the photopic spectrum. The curve under
discussion is shown in Fig. 2 of Chapter 2, and of the two there
shown that labelled "photopic" is the one relevant to our
present purpose. As noted above, it demonstrates that the
spectral extremes are less bright than the greens, yellows and
oranges of the centre. These data must clearly find some place in
any adequate theory of colour vision.

We have noted above certain facts concerning two-component
mixtures. Thus, such a mixture will yield psychological primary
red, a series of extra-spectral purples, or a complementary
achromatic stimulus. In the latter case, if the intensities of the
two components are not correctly adjusted, the result is an unsaturated stimulus whose hue is that of the component present in
too great a proportion. All these data may be incorporated in the

---

[1] This figure is for the spectrum of Abbot-Priest Sunlight (Priest, 1918).

so-called Colour Triangle (*e.g.* Parsons, 1924), whilst the achromatic stimulus relationships may be incorporated with these in the tri-dimensional Psychological Colour Solid (*e.g.* Troland, 1934). These graphic methods of illustrating the data of vision have, of themselves, little significance, save as an aid to schematizing the various interrelationships which they represent. To permit the elaboration of a theory of colour vision, the relevant data must be presented in a somewhat more quantitative form.

The most important of these to be considered are the results of calibrating the spectrum in terms of three primaries, thus yielding essential data as to the results of three-component mixture, as opposed to two-component. Attempts to do this date from early times (Newton, 1794) almost to the present day (Guild, 1931). On at least one point, however, all investigators are agreed. With a mixture of two primaries selected from the opposite spectral extremes and one from the centre, it is possible to match all spectral hues (and indeed all extra-spectral ones), and to cover a wide range of saturations, by variations of the intensity ratios of the three components only. The intensity of one or even two members of the three can, of course, be made zero, yielding either the two-component system discussed above, or a homogeneous stimulus. Thus, it must not be thought that all spectral hues can be obtained by colour mixture; this is, indeed, most certainly not the case. Neither red nor green can be obtained in any satisfactory degree of saturation whatever by colour mixture. To say this, however, is only to make certain reservations with regard to the wave-lengths of the monochromatic primaries selected; if one of these is chosen from the green region of the spectrum, and one from the red end, these facts can be allowed for; it is then necessary only to reduce the other two components nearly to or completely to zero to obtain the two hues in question.

Probably the most satisfactory figures for the results of calibrating the spectrum in this way are those of Wright (1928, 1929). These are illustrated in Fig. 7, in one of the many forms in which they may be presented. Wright's primaries were fixed at 460 mμ. (blue), 530 mμ. (green) and 650 mμ. (red), and it will at once be observed that negative values must be allowed for on the luminosity scale. This state of affairs is inseparable from any attempt to calibrate the spectrum in terms of real, mono-chromatic primaries. The curves of Fig. 7 can, however, be made to yield equations connecting the three primaries and the resultant hue, and such equations may be solved for other primaries, either real, monochromatic ones, as in Wright's case, or for primaries themselves composed of more than a single monochromatic stimulus; or indeed for primaries which have no

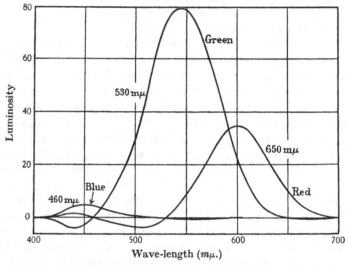

Fig. 7. Colour mixture data for the standard spectrum, incorporating the photopic luminosity data (after Wright).

existence in or out of the spectrum. Under these latter con-ditions, the primaries are said to be unreal or imaginary, and their appropriate choice will enable the avoidance of negative luminosity values.

The data shown in Fig. 7 incorporate the results of measure-ments of the brightness of the spectrum for the photopic eye. That is to say, when summed they will yield a curve similar to the photopic one of Fig. 2. It is, however, possible to divorce these two sets of data, and present those for colour mixture alone. Thus, the height of the three primary curves may be arbitrarily made the same at their maximum values, or, alternatively, the areas under each curve may, by multiplying each by a different factor, be made arbitrarily equal. Either of these transformations, and indeed many others, may be made with curves for real or imaginary primaries, a state of affairs accounting for the vast apparent dissimilarity between the data presented in different places. Such differences should not be permitted to obscure what is perhaps the most fundamental fact of colour vision, namely, that the whole visible spectrum may be represented in terms of three primaries, appropriate mixture of which will match any single hue.

This basic fact was, as we have noted above, early discovered. It is at once evident that it must be incorporated in, indeed should, perhaps, form a fundamental part of, any theory of hue perception. The necessity for this conclusion was first observed in 1807 by Young, who propounded a theory supposing that the eye contained three kinds of light-sensitive receptor, each maximally sensitive to, and producing a characteristic sensation of red, green or blue. This extremely important suggestion was not noticed either by the biologists or the physicists of his time, but was first adopted by Maxwell almost fifty years later (Maxwell, 1855, 1890) and subsequently by Helmholtz (1866, 1924). Helmholtz elaborated Young's original idea to a considerable extent, and the theory based upon it is now ordinarily known as the Young-Helmholtz Trichromatic Theory. It has more recently been further elaborated and modified by, notably, König (1903, pp. 88–107) and by von Kries (1905). Into the various minor modifications of the original supposition proposed by these writers and others it is not necessary to go here. We may note simply that the trichromatic concept is one which provides a readily understandable account of the nature of the peripheral process, beyond which it makes no claims to go. This point has not always been clearly understood in the past, and as a result objections irrelevant to the theory in question have frequently been raised.

On the trichromatic formulation, a sensation of yellow results from the simultaneous stimulation in appropriate degree of two receptors, the "red" and the "green". Similarly, a sensation of white results from the appropriate degree of simultaneous stimulation of all three receptor types.[1] This conclusion is necessary to Young's theory, and to every modification of it which has been proposed. It is on these two points, however, that most dispute has arisen. It has been argued, notably by Hering (1888, 1920), Ladd-Franklin (1893, 1929) and G. E. Müller (1896), that yellow and white are unique sensations, and cannot be imagined as arising from the simultaneous stimulation of two or three receptor types. Thus they point out that mixtures of red and blue light yield a series of extra-spectral purples in which either component is readily distinguishable. They are, therefore, unable to accept the trichromatic theory in its deductions from the fact that yellow, a sensation unique in itself, results from a mixture of red and green. The same is true for complementary mixtures which together yield a colourless sensation, somewhat euphemistically called white. Hering, Ladd-Franklin, Müller and others have advanced theories of colour vision postulating separate receptor systems for yellow and white. Of these, the most widely held is the Opponent Colours Theory of Hering. According to this, there are six "fundamental sensations", red, green, yellow, blue, white and black. These are divided into three complementary pairs, one hue of each pair causing a reaction to proceed in one direction, and the other the same reaction in the opposite. Thus, black, green and blue yield assimilatory changes (anabolism), whilst white, red and yellow cause dissimilatory changes (katabolism) (Rivers, 1900, p. 1112). It is of great importance to note that the substances in which these changes take place are not necessarily located in the retina, as is often erroneously stated. They are held to "exist somewhere in the sub-cortical visual paths: their exact position is not defined,..." (Parsons, 1924, p. 298).

It would be out of place here to attempt any summary of Ladd-Franklin's or of Müller's theories of colour vision, for these

[1] We are here speaking of cone-mediated vision. Scotopic vision is essentially achromatic, and thus it is possible, on the Duplicity Theory, to regard the rods as constituting a separate "white" receptor system. This will be of significance only at low illuminations below the cone threshold, however.

have never received the general support accorded to the trichromatic formulation and, to a less degree, to the rival opponent colours theory. It may, however, be of use to point to the existence of the excellent summaries of these and other theories, and of various further modifications of the trichromatic concept not mentioned above, to be found in Parsons' *Colour Vision* (1924, Section v, pp. 284–314).

Many efforts have been made to differentiate the two major theories of hue perception. These have, in the majority, centred round the problem of the uniqueness of the yellow sensation. It is clearly evident that the question to be settled is, simply, whether or not there is a separate receptor whose stimulation is characterised by this sensation, or whether it arises from the simultaneous stimulation of two separate systems. This matter cannot be satisfactorily settled for a single eye, for it is never possible to assert with confidence whether one or more receptor types are involved. From the earliest times, therefore, efforts have been made to secure the yellow sensation by simultaneous stimulation of the opposite eyes with red and green monochromatic light. The first important attempt to secure the binocular fusion of colours in this fashion was that of Helmholtz (1909–11), who, however, failed to secure the necessary conditions for success. This fact to some extent discouraged further experiments along these lines, though Hering himself performed the fusion successfully. As a result of the work of Trendelenburg (1923) and Rochat (1925) there is now no doubt, however, that binocular fusion of red and green stimuli will yield a yellow sensation. Hecht (1928) has described a simple apparatus with which the experiment may be convincingly performed by a majority of observers. Thus stimuli of wave-length and intensity suitably chosen from the red and green regions of the spectrum will, if permitted to act upon opposite eyes, yield a sensation of yellow. Further, if yellow, now indicated to be the product of simultaneous stimulation of two receptors, is permitted to act upon one eye whilst a selected blue falls upon the other a colourless sensation can, under appropriate conditions, be secured. In this latter experiment, it is generally assumed that three receptor types are operative.

These data provide striking evidence for the Young-Helmholtz formulation. Indeed, it is difficult to feel that there can still be any necessity for entertaining an opposing view, as to the broad outline of the peripheral process. We may now, however, briefly review the remainder of the evidence for the trichromatic theory, together with that in favour of the opponent colours formulation. No attempt will here be made to perform this laborious task in detail, for this has been done adequately in many summaries in the past (see especially Parsons, 1924).

In addition to the colour mixture material and that on the binocular fusion of colours, it is customary to cite, as bearing upon the trichromatic concept, data derived from the study of so-called colour blindness. Thus in certain types of this abnormality it is possible to match all hues of the visible spectrum with two, rather than three, primaries. This condition is ordinarily described as *dichromatic* vision, and is considered to be due to the absence of, or failure of correct function of, one of the receptor types. Dichromats may be divided into two types, *protanopes* and *deuteranopes*, the principal difference between the two being in the wave-length requirements of the primaries with which such persons match the spectrum. Protanopes are insensitive at the long wave-length end of the spectrum, whilst deuteranopes are insensitive in the middle of the spectral scale. All dichromats can employ two primaries if these are of suitably selected yellow and blue light. The protanope will, however, succeed equally well with green and blue primaries, whilst the deuteranope will require red and blue. It should be emphasised that matches valid for the trichromat, or normal eye, are valid also for the dichromat (*e.g.* Parsons, 1924, p. 181). Dichromatic vision may, therefore, be derived from normal by the simple loss of one hue-perceiving mechanism. It is evident that Hering's Opponent Colours Theory can offer no such clear explanation of these facts.

Of the remaining forms of colour deficiency, a further form of dichromatism may first be mentioned. This state, which is extremely rare, is known as *tritanopia*, and it is held that yellow and blue psychological primaries are simultaneously absent for such individuals. It is evident at once that the simultaneous disappearance of yellow and blue provides some argument in favour of the opponent colours theory, though the rarity of the disease makes it difficult to assess the importance of this evidence with certainty.

The trichromatic theory is unable satisfactorily to explain that form of colour disturbance known as *anomalous trichromatism*. Persons with this maladjustment require three primaries as does the normal trichromat, but their ratios for any given spectral match are altogether different from the normal.

Monochromatism, or complete colour blindness, is a state characterised by complete absence of all perception of hue. The condition is congenital, and is sometimes known as *achromatopia*. It is usually associated with certain further defects of vision, such as myopia, photophobia and nystagmus. A good summary of the known data is given by Parsons (1924, pp. 197–201). The scotopic visual acuity approximates to normal, and the spectral luminosity curve, even at high overall intensities, is found to agree closely with the normal scotopic luminosity curve. The condition is usually ascribed to complete functional absence of cones (Parsons, 1924, p. 220), and has, therefore, more relevance to the Duplicity than to the Trichromatic Theory.

For a more detailed description of the phenomena of colour blindness, reference may be made to the classic work of G. E. Müller (1924). We may here note that the original suggestion of Young, that colour blindness was due to loss of function of one or more types of colour-sensitive receptor, is inadequate to explain more than a few of the many abnormal states to be found. The further supposition that, under certain circumstances, one colour-receiving mechanism may be transformed into another enables the trichromatic concept to cover more of the relevant data, but certain weaknesses are still apparent in the theory thus produced. König (1903) first pointed out that the luminosity of the spectrum for a deuteranope is not significantly different from the normal; though for the protanope great differences are apparent, the maximum brightness being shifted towards the short-wave end of the spectrum, and the luminosity of the long-wave end being greatly subnormal. It is evident that neither loss of function nor transformation of one receptor group into another will account for both these phenomena at once, unless it be assumed that the green sensitive receptor plays a very minor role in determining the luminosity curve.

Evidence which is usually held to argue in favour of the Opponent Colours theory derives from the study of the perception of colours in the periphery of the visual field. If a small coloured object is moved in from the extreme periphery towards the centre, a point will be reached at which its hue can first be recognised. Proceeding in this manner at various angles, it is possible to plot the field of vision for several colours, though the brightness and area of the test objects must be rigidly controlled. Using spectral stimuli, Abney (1913) has plotted the visual field in this way, at various illuminations. From his work it may be seen that the generally accepted view, namely that complementary red and green cease to be appreciated at the same visual angle, cannot be substantiated under all conditions. Indeed, it would appear in general that red ceases to be appreciated as such very much sooner than does the complementary green, and that green and blue disappear at approximately the same angle (Fig. 8). Thus Hering's prediction that vision for the complementaries should cease at the same points, owing to loss of red-green "substance" and of yellow-blue "substance" before the loss of the corresponding white-black mechanism, has received little support from this work.

Abney's results are somewhat difficult of interpretation on the trichromatic theory. The fact that yellow can be seen more peripherally than red and green (Fig. 8) is difficult to reconcile with the fact that such a sensation supposedly results from simultaneous stimulation of "red" and "green" receptors. Using other spectral choices, moreover, Abney has shown that green may cease to be seen most centrally, red much farther in the periphery, with yellow following immediately. This can be interpreted as indicating that a sensation of yellow requires a fully operative "red" receptor system, but that the "green" is less important; this is scarcely in strict accord with the trichromatic view.

More important evidence favouring the opponent colours view than that outlined above is to be found in the phenomena of

contrast colours. Thus if a piece of grey paper is placed on a
large red square, the grey appears tinged with complementary
green. To this and similar phenomena the term *simultaneous*

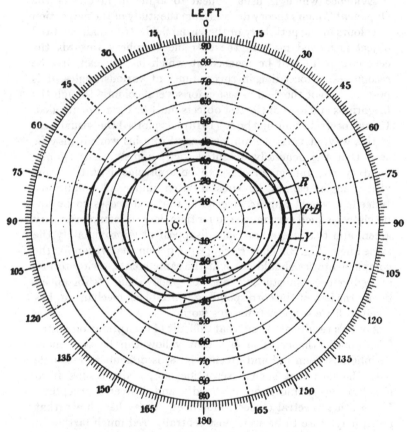

Fig. 8. The fields of vision for complementary colours (after Abney).

contrast is ordinarily given. If a coloured square is fixated for
some seconds and the eyes then turned on to a grey background,
a complementary coloured square, or *negative after-image*, may
on occasions be observed, constituting an example of the
phenomenon of *successive contrast*. These effects may be ac-

counted for on Hering's theory by supposing that when, for example, anabolism is induced in one area, autonomous katabolism is induced in neighbouring areas, and (for successive contrast) subsequently in the same area as that stimulated. Hering's explanation of the positive after-image phenomena is less satisfactory. Thus, if a very dark background is fixated after viewing a red object, a positive red after-image is seen. This was ascribed by Hering to exhaustion of the anabolic process, leaving only feeble autonomous katabolic changes. The difficulty of this interpretation arises from the fact that the hue of a positive after-image changes to the negative complementary on fixation of a surface brighter than the after-image itself. Though the theory of opponent colours offers an easily comprehensible explanation of the major facts of contrast and induction, McDougall (1901) points out that there are many similar effects which this theory cannot account for. *Inter alia*, he notes that Hering's explanation of the positive after-image phenomena is in general difficult of acceptance.

Helmholtz himself has attempted to account for the facts of induction on the trichromatic theory. Very great difficulty attaches, however, to acceptance of the theories he puts forward. It is possible to account for the negative after-image by supposing that the previous stimulation has produced some selective fatigue. The light remitted from the background upon which the after-image is projected acts upon all three receptor mechanisms; these, however, owing to previous stimulation, are less sensitive than they would otherwise be. Moreover, that maximally sensitive to the colour of the object previously viewed is more greatly affected than are the remaining pair. It is not difficult to see, therefore, that if light acts upon all three systems, then the *relatively* increased response to the complementary can be explained as yielding the negative after-image.

This is the view adopted by Helmholtz, though it leads immediately to many difficulties. The excellent evidence derived from the study of dichromatic vision by v. Kries (for summary see Parsons, 1924, pp. 174–92) and by Abney (1913) shows that stimuli of greater wave-length than 650 mμ. (approximately) do not act at all on the blue-sensitive component. It is thus extremely difficult to explain after-images from the yellow-blue complementaries, or even the increased saturation of yellow found after previous stimulation with the complementary blue; for in theory the yellow light should act only on the red and green components, and should therefore be unaffected by the state of sensitivity of the remaining one. As Parsons says (1924, p. 231): "We must, therefore, accept the theory as explaining satisfactorily *either* the phenomena of after images *or* those of dichromatic vision, but not both." (Italics in original.)

The positive after-image phenomena (*successive induction*) have been explained by Helmholtz as being due to residual excitation—to nervous activity persisting, that is to say, after removal of the visual stimulus.

Simultaneous induction Helmholtz makes no attempt to explain. The observed phenomena, he believes, can only be interpreted as constituting illusions of judgement, an argument difficult of acceptance in view of the apparent kinship between the phenomena of simultaneous induction and those of successive induction.

The facts of simultaneous contrast can only be explained on the trichromatic theory with some extension. We may, for example, account for them by assuming a certain spatial spread of selective fatigue. For the existence of such an effect in achromatic vision (*i.e.* vision for white light) there is considerable evidence (Lythgoe and Tansley, 1929 *b*).

McDougall (1901) believes that Helmholtz's expansion of Young's trichromatic theory to cover the facts of induction is so unjustifiable as to render the description "Young's Theory" preferable to the more conventional "Young-Helmholtz Theory". Whilst supporting the three-components formulation, therefore, McDougall has himself advanced an extension of this to cover the contrast phenomena. This writer also combines the trichromatic and the duplicity theories, for he finds it necessary to postulate an additional "white-sensitive" receptor system, which he thinks to be located in the retinal rods. With this view few modern theorists would quarrel, though the white sensation obtained by foveal stimulation cannot be accounted for in this way.

With such necessary extensions as these it would appear, from the brief summary given above—as indeed from a fuller exploration of the original literature itself—that the data of hue perception support the trichromatic concept to the exclusion of opposing views. Clearly, such a theory cannot give a complete account of all the phenomena of chromatic vision; as a description of the peripheral mechanism, however, it seems preferable to such other hypotheses as have been advanced. We may now turn, therefore, to a consideration of the suggestions which have been made in regard to the problem of specifying the spectral sensibilities of the three cone types.

When the data of colour-mixture experiments are adjusted to give the photopic visibility curve by addition (as in Fig. 7), or when the original experiment is so performed that the primaries are chosen and adjusted so as to match each point of an actual solar spectrum in intensity as well as in hue, the resultant curves may well indicate the relative sensitivity to varying wavelengths of the three receptor types. Unfortunately, however, there is no *a priori* reason to suppose one choice of primaries to be superior to another, and, as we have noted above, a very large primary range is possible. It is thus necessary that additional data be brought to bear upon the problem, to aid the selection of primary wave-lengths.

The data derived from the study of deuteranopia and of protanopia, which we have discussed above, can offer some assistance here, for on the trichromatic theory it may be assumed that the "red" and "blue" curves should incorporate the stimulus mixture relationships for the former, and the "green" and "blue" curves those for the latter. Now in curves of the general type shown in Fig. 7, which, when summed, give the normal photopic luminosity curve, the areas under the curves for the three primaries are by no means equal. This will be the case irrespective of the primary choices. The inequality of area may be removed by multiplying the data from which each curve is derived by a different factor, thus equalising the areas whilst keeping the form of each curve constant. This modification, together with the recomputation in terms of other primaries required to avoid negative values on the luminosity scale and to satisfy the colour-blindness data which we have just discussed, was first accomplished by König and Dieterici (1892); to the final curves thus produced they attached the term basic sensation curves (*Grundempfindungen*), for it was suggested that they might represent the spectral sensibilities of the three cone types (Fig. 9). Owing to the equalisation of areas of the three curves, however, the photopic luminosity curve cannot be derived from these curves by addition, and, for this and other reasons, they have been rejected by many authors. Thus in a paper published shortly after their first description by König and Dieterici, Helmholtz (1891) abandoned them owing to their shortcomings

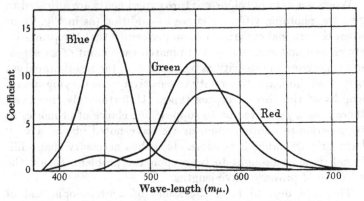

Fig. 9. *Grundempfindungen* Curves of König and Dieterici.

in these respects. This fact does not appear, however, to have prevented the *Grundempfindungen* achieving what Hecht (1934, p. 798) has well termed "an air of great respectability" which they can scarcely be thought to deserve. Hecht (1930) has adduced a formidable array of facts mitigating against their acceptance as spectral sensitivity curves for the three-cone types.

It should not be thought from the foregoing that the omission of the area equalisation process will yield curves which satisfactorily fulfil the purpose we now have in mind. It has been noted above that great difficulty attaches to a correct choice of primaries, real or imaginary, for no experiment has as yet been devised which differentiates with certainty between the possible choices in this respect. The dichromatism data yield some evidence here, as we have seen, but scarcely sufficient to preclude consideration of other relevant matter. Thus it might be expected that wave-length discrimination would be most acute at the crossing point of the primary receptor curves. As a minor variant on this theme, it might be supposed that these crossing points should represent the three "psychological" primaries of the

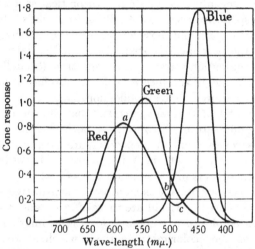

Fig. 10. "Cone response" curves (after Purdy). The following criteria are satisfied: (1) The crossing points *a*, *b*, and *c* correspond to the yellow, green and blue psychological primaries for a normal observer. (2) The green and blue curves represent the stimulus mixture relationships for the protanope. (3) The red and blue curves represent the stimulus mixture relationships for the deuteranope.

spectrum, yellow, green and blue. This latter view is that adopted, for example, by Purdy (1935). He has, therefore, computed the stimulus mixture relationships for the 1931 standard observer of the International Commission on Illumination so as to fulfil these conditions, together with certain unspecified data in regard to deuteranopia and protanopia (Fig. 10). He thus shows curves extremely similar to the basic sensation curves of König and Dieterici, though labelling the ordinate scale "Cone Response". This will be seen on comparing Figs. 9

and 10. It is evident, therefore, that "the air of great respectability" acquired by the *Grundempfindungen* has not yet altogether evaporated. Precisely the same strictures apply, however,
as they did to König and Dieterici's curves.

These matters have been well discussed by Hecht, who has,
moreover, made an interesting attempt to derive spectral sensibility curves for the three components which avoid these pitfalls.
He has attempted also to embrace more data than have ordinarily been encompassed in similar theoretical schemes of the
past (see especially Hecht, 1934). Thus, in Hecht's scheme, the
colour-mixture data and those for the brightness of the spectrum
are simultaneously incorporated. Additionally, data on wave-
length discrimination throughout the spectrum (Steindler, 1906),
and on saturation of the spectral colours, also find a place. This
latter point requires some amplification.

It is a matter of general agreement that the spectral colours
are not equally saturated, as we have noticed earlier. Thus the
spectral extremes seem, in general, more saturated than the
yellows and oranges of the mid-spectrum. In order to incorporate these facts into a scheme such as that of Hecht, however, it is necessary to have them in some quantitative form. The
measurements used by Hecht are those of Priest and Brickwedde
(1926) who employed a technique for the estimation of what
they have termed "colorimetric purity", which, as we have seen
(p. 44), is defined as being inversely proportional to the
saturation at any given wave-length. As we have noted, Priest
and Brickwedde determined the minimal amount of each
spectral hue which had to be added to a white field to make it
just perceptibly coloured. As the hue was added, white light was
subtracted, in order to ensure brightness compensation. To
justify this technique, it is essential to assume that the actual
amount of "fully saturated" colour which has to be added to
reach threshold value is identical for all wave-lengths, and from
this it follows that the saturation of the hue actually employed
is inversely proportional to the amount of it which has to be
added to the constant white field for its addition just to be
noticed. This does not seem in any way an essential conclusion,
and we shall return to this matter again.

Hecht's spectral sensibility curves (Fig. 11) differ markedly from those discussed above in that they show the sensibilities of the three-cone types to be closely similar. On general biological grounds some correspondence would appear likely, but certain obvious weaknesses in Hecht's formulation cannot be escaped. Thus, in order to describe *all* the data outlined above it has been necessary to plot three sets of curves, called by Hecht V_0, G_0, and R_0, V_0', G_0', and R_0', and V_L, G_L, and R_L. Of these the first set are held to describe colour mixture and photopic luminosity perfectly, and hue discrimination and saturation approximately. The second set describes saturation with especial accuracy, whilst the third does fullest justice to wave-length discrimination.

A second weakness of this formulation is also apparent, in that the deuteranopic and protanopic colour-mixture data are not here incorporated. The photopic luminosity curve is approximately yielded by addition in all cases, however. A third, and very serious weakness lies in the fact that the differences between the three sets of curves required to do justice to the specific data with which they deal are considerably greater than the differences between the three receptor curves themselves. One is therefore forced to assume either that the data which they describe are hopelessly inaccurate, despite their careful selection, or that the significance of the data themselves has been in some fashion misinterpreted. Close inspection of the final curves (Fig. 11) reveals, it is believed, not only that the latter rather than the former is the case but also the nature of the misinterpretation in question.

All the curves, as we have noted, are partly derived, either accurately or approximately, from the saturation of the spectral colours data. These have in turn been derived from the colorimetric purity measurements of Priest and Brickwedde, and are accurately incorporated in the set of curves V_0', G_0', R_0'. Inspection of these reveals that they depart more radically from the remaining two sets than do the latter *inter se*. This immediately renders the saturation data suspect, and we may therefore return to a consideration of Priest and Brickwedde's procedure.

I have indicated above that their results are suspect in regard

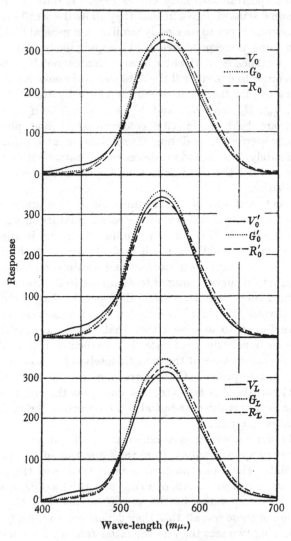

Fig. 11. Hecht's three sets of cone primary curves.

to their assumption that an equal amount of theoretically "fully saturated" colour must be added just to colour the white field, irrespective of wave-length. This assumption may well be unfounded. Thus it is difficult to escape the conclusion that yellow, even if it could be obtained fully saturated, would be more nearly identical with white than, for example, violet. The fact that it is probably the product of stimulation of two receptors adds weight to this assumption, for it cannot be overlooked that appropriate stimulation of all three receptors can yield an entirely colourless sensation. In view of its highly speculative nature, this supposition cannot be further laboured, but it is perhaps worthy of note that it points immediately to the solution of the difficulties outlined earlier. The approximate form of all Hecht's curves is governed to a not inconsiderable extent by Priest and Brickwedde's data, and it would seem that, if these are omitted (in view of their theoretical insecurity), a more likely set of spectral sensitivity curves is obtained. Not only is the resemblance between the three individual curves thus rendered less marked, a weakness of Hecht's formulation too obvious to need further stressing, but the dichromatism data can be brought into closer line than is otherwise the case. Certain other advantages, into which we need not go here, are also secured.

A recent formulation of the three-component spectral sensitivity curves, in which the saturation data find no place, is that of Wright (1934), reproduced in Fig. 12. Derived from work on colour adaptation phenomena rather than by direct measurement, though otherwise treated in essentially the same fashion as König and Dieterici's *Grundempfindungen*, Wright has yet omitted to equalise the areas beneath the three curves. Consequently, on addition they give the photopic luminosity curve, a very desirable feature as we have noted above. One further important difference between these curves and König's may be noted, in that it will be seen that the "green" curve descends below 0 on the ordinate axis, thus becoming negative. As Wright says (1934, p. 69): "The physiological mechanism by which such an effect could be produced cannot be visualised very readily, but it would apparently necessitate the assumption that all three fundamental responses have some quality in common, so that one response could produce a subtractive effect on another. This quality must probably be in the nature of an inherent 'whiteness', and it is on an assumption of this sort that saturation differences might be explained." Thus Wright's curves appear attractive in that they account for the saturation effects qualitatively, by having negative luminosity values; for the photopic luminosity curve precisely; and also show the blue receptor failing to respond beyond 530 mμ. or so, in accordance with the data obtained for a dichromatic eye which we have discussed above (p. 51). It must be again stressed, however, that this latter consideration

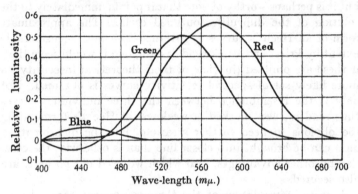

Fig. 12. Cone response curves of Wright (1934).

precludes an adequate explanation of the increased saturation of yellow after blue stimulation. As we have already noted, however, one cannot reconcile these two sets of data at once, and the dichromatism material at present appears the more important. It may fairly be said that, bearing its limitations in mind, Wright's formulation is at least as attractive as any so far proposed, and considerably more so than many.

The elaboration of further spectral sensitivity curves should, in the present author's opinion, await further experimental evidence. Whence this will come we cannot say with confidence at present. In this connection it is, however, of interest to note the recent highly indicative work of Stiles (1937), on a colour effect obtained when the point of entry of a monochromatic ray is moved across the pupil. In Stiles' experiment two points of monochromatic light are simultaneously brought into focus on the anterior corneal surface. As one of these is moved across the pupil its hue changes by an amount which may be measured by making the second and stationary point adjustable for wavelength, thus enabling a match to be made. The results obtained in this fashion show that the hue changes in such a manner that the apparent wave-length becomes longer as the stimulus moves from a point approximately in the centre of the pupil to the periphery, by an amount which varies slightly according as the departure is to the temporal or nasal sides. This interesting phenomenon can most adequately be explained on the supposition that the three sets of cones contain photochemical substances with different spectral absorption curves, and are also differentiated for size. Its existence provides some evidence, therefore, for the reality of the trichromatic view. We may well feel that further speculation as to the differences of spectral sensitivity in question should await the accumulation of additional matter in this and other ways.

More especially we may hope that the electrophysiological technique, whose results have been so highly indicative in other fields, may soon be developed towards the solution of this problem. The satisfactory interrelation of the vast mass of facts on human colour perception remains still the central question of sensory physiology, though its solution on the basis of the information we have at present has been attempted by great men and by lesser. The too evident fact that all have so far failed indicates the necessity for further data and new ideas.

SECTION II

AUDITION

CHAPTER 5

The Anatomy of the Ear,[1] and the Function of the Peripheral Mechanism

In accordance with obvious anatomical features, and to facilitate a description of its somewhat complex spatial relations, the human ear may conveniently be divided into three parts. Of these the *outer ear* (Fig. 13) consists of the *pinna* or auricle, together with a convoluted tube, the *external meatus*, leading at its inward end to the *tympanic membrane*. This is often loosely called the *tympanum* or ear drum, but these latter terms should preferably be used to include other parts of the ear, in addition to the drum-skin itself. The latter is a diaphragm of connective tissue, covered externally by modified skin continuous with that of the meatus, and internally by a mucous membrane, as is the remainder of the *middle ear*. In shape, the tympanic membrane is conical, and it is tightly stretched and translucent in the normal ear.

It has on occasions been argued that the outer ear functions as a "collector" of sound waves, in some fashion reflecting these up the meatus. It is at once clear, however, that such reflection can only take place from a surface which is large in comparison to the wave-length of the sound which it is designed to reflect. In the human ear this cannot be true except for the highest audible frequencies. The cavity in the pinna into which the meatus leads, the *concha*, is, however, believed to increase the ear's sensitivity by resonance over a small band of frequencies (Beatty, 1932, p. 2).

At the apex of the tympanic membrane, on its interior surface, is the *malleus*, which, with the *incus* and *stapes*, comprise the ossicles of the middle ear. These are small bony structures

---

[1] As in Chapter 1, this anatomical description is not intended to be comprehensive, but rather to provide such a brief summary as will facilitate the discussion which is to follow. Quain's *Anatomy* (1909) offers a good and full account.

serving to transmit the vibrations of the tympanic membrane to the oval window (or *fenestra ovalis*) of the *cochlea*. The latter is that part of the *inner ear* concerned with the sense of hearing. Under normal conditions, the middle ear has no communication with the outer air, though it is itself air-filled. Large variations in pressure of the external air are prevented from distorting or injuring the tympanic membrane, however, by the provision of

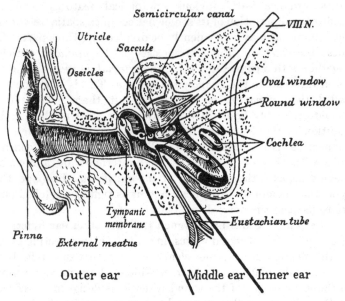

Fig. 13. Diagram of the human ear.

the *Eustachian Tube*, leading from the middle ear to the naso-pharynx. At its lower end this is closed by a valve actuated by the *tensor palati* muscle. The act of swallowing opens this momentarily, and thus releases any negative or positive pressure which may have developed in the middle ear cavity.

Movements of the malleus are communicated to the incus through a saddle-shaped articulation between the two, whilst these in turn reach the stapes through the *incudo-stapedial junction*, which is to some extent reminiscent of a ball-and-socket joint. The stapes foot-plate fits into the oval window of the cochlea, to which it is attached by an annular ligament. In addition to the ossicles, the middle ear contains two muscles. The larger of these, the *tensor tympani*, is attached to the malleus in such a way that its resting tonus serves to maintain the tympanic membrane in a condition of tautness. In the event of this being relaxed, the efficiency of the membrane mechanism falls somewhat. The second of the middle ear muscles is the *stapedius*, which, as its name implies, is attached to the stapes. It is extremely minute, and is held to be the smallest muscle in the human body. One of its functions is thought to be the maintenance of satisfactory articulation of the ossicles (Hartridge, 1934a, p. 926).

The primary function of the middle ear mechanism is to transmit vibrations, communicated from the outer air to the tympanic membrane, to the fluid-filled cochlea. The ossicles are so arranged that an inward movement of the tympanic membrane results in an inward movement of the oval window. It is to be noted, however, that the stapes foot-plate is so suspended against the oval window that its overall amplitude of movement is approximately one-third of that of the centre of the tympanic membrane. It is evident, therefore, that the ossicles form a lever which, among other functions, reduces the amplitude of movement of the oval window. It has been noted by Beatty (1932) that this mechanism does not, as has sometimes been argued, reduce the efficiency of the ear, but in fact actually increases it. Thus, when a light, air-driven diaphragm such as the tympanic membrane must drive a device offering a comparatively large resistance such as the fluid damped oval window a situation arises, familiar to every acoustical engineer, in which two dissimilar impedances must be matched. The necessity for such matching occurs frequently in communication engineering, when dissimilar electrical impedances have to be matched by the interpolation of a transformer between them. The now old-fashioned acoustical recording device presented a similar problem in the mechanical sphere. Here, a light diaphragm at the end of the recording horn was used to drive the cutting stylus against the resistance offered by the recording material. As in the ear, a lever accomplishing a reduction of amplitude was employed, the ratios being adjusted until matching was obtained. In the ear, the reduction ratio of 3 to 1 would appear, as far as may be said, to be appropriate to "match" the tympanic membrane to the oval window and the cochlear fluids.

Any vibrating diaphragm must possess one or more resonant periods; that is to say, its response to certain frequencies will tend to be greater than elsewhere, owing to enhancement by the phenomenon of resonance. It has been pointed out that the shape and structure of the tympanic membrane is such as to render it approximately free from such defects, that is to say aperiodic (*e.g.* Hartridge, 1934*a*, p. 925). It is to be noted, however, that the asymmetrical arrangement of the membrane and ossicles necessarily involves that the phenomenon of rectification by the mechanical system of the ear should be present. This fact is responsible for the summation and difference tones heard when two pure tones of dissimilar frequency fall upon the ear at one time (Barton, 1922, p. 391). Thus, if tones of 200 and 500 c.p.s.[1] are simultaneously sounded, tones of 300 (difference) and 700 (summation) c.p.s. should be heard in addition to these. In practice, the summation tone is somewhat difficult to hear, owing to the masking influence of the two lower objective tones, though the difference tone will be readily audible to a majority of observers.

A still further disadvantage of the arrangement of the mechanical system of the ear is that it is responsible for the introduction of so-called subjective harmonics. Thus, if a pure tone is permitted to act upon the ear, it is ordinarily accepted that the vibrations of the cochlear fluids will be of such wave form as to contain, in addition to the fundamental tone itself, decreasingly small powers of all its harmonics from the second upwards. This effect, however, is of small importance except for intense stimuli, and, moreover, when these are themselves extremely pure, for at a normal intensity level the subjective harmonic intensity is itself very small.

The remaining properties of the middle ear are those due to the presence of the tensor tympani and stapedius muscles. It has

---

[1] Cycles per second.

To face p. 73

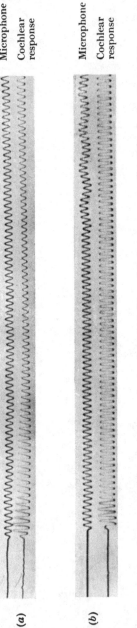

Microphone

Cochlear
response

Microphone

Cochlear
response

(a)

(b)

Fig. 14. (a) Oscillogram showing reduction in amplitude of the cochlear response, produced by contraction of the *tensor tympani* and *stapedius* muscles. Upper record shows constancy of sound stimulus, as measured by microphone. Frequency of stimulus: 1000 c.p.s. (b) Control experiment demonstrates abolition of effect, following paralysis of muscles by section of motor nerves. Frequency and other conditions exactly as in (a).

been noted earlier that the normal tonus of these is responsible for the maintenance of tautness of the tympanic membrane and of satisfactory articulation of the ossicles, respectively. It is now known, however, that in certain animals (Kato, 1913) and some human beings (Köhler, 1909) the muscles exhibit in addition a reflex response to sound incident on the ear. Thus it has been observed that an auditory stimulus, at least if of adequate intensity, will cause both muscles to contract. As to the effect of these contractions various views have been held. The majority of investigators have believed that, by increasing the stiffness and consequently reducing the amplitude of movement of the vibrating mechanism, the muscles exercise a protective function in that they. prevent movements of excessive amplitude in response to stimuli of high intensity. The opposing view that the sensitivity and therefore the amplitude of movement was *increased* during reflex contraction has, however, also been held. This divergence of opinion was in large part due to the difficulty of measuring the amplitude of movement of the stapes foot-plate or of the internal cochlear structures, and of the effect of the contracting muscles on this. It is now known, however, that the cochlea will yield an alternating electrical potential in response to auditory stimuli, and that the size of this is governed by the stimulating intensity (for a full discussion of this effect see later). This so-called "cochlear effect" has been used to study the function of the middle ear muscles. Thus it has been shown that, in the cat, the amplitude of the electrical effect, and therefore it is assumed of the vibrations of the cochlear members, is reduced by contraction of the two muscles, and that this reduction is abolished if the muscles are paralysed (Fig. 14a, b) (Hallpike and Rawdon-Smith, 1934a). It is evident, therefore, that in this animal the muscles have a protective action in that possible damage to the delicate inner ear contents by loud sounds is prevented by reflex muscular contraction and consequent lowering of sensitivity of the whole receptor. This mechanism may be compared with the "automatic volume control" common to the majority of present-day radio receivers, whereby the sensitivity of the receiver is adjusted automatically by the amplitude of the carrier wave of the station to which it is tuned. In this way, gross overloading on nearby stations is avoided, though the maximum sensitivity of the receiver may be high.

Two further points should be made in connection with the middle ear muscles. First, it can be shown in the cat that the sound applied to one ear only will induce an obvious contraction in the muscles of the opposite. This phenomenon may be compared exactly with the Consensual Pupillary Reflex discussed in Chapter 1. Secondly, the latent period for each muscle may be separately measured by recording the electrical myogram, simultaneously obtaining a record from a microphone placed at an equal distance from the sound source. This may be done for ipse-lateral and for contra-lateral stimulation. The results obtained show that, for a reasonably intense stimulus, the stapedius latent period is approximately 6 milliseconds, whilst the tensor tympani shows the slightly greater figure of 7–8 milliseconds. In each case the contra-lateral figure is slightly greater than the ipse-lateral, as might be expected.[1] It is evident that these time intervals are of such length that the muscles can offer no protection to the ear from loud sounds of extremely short duration, such as nearby gun-fire.

The inner ear, often known as the labyrinth, is contained in a cavity in the petrous portion of the temporal bone. The cavity, or *osseous labyrinth*, is of somewhat complicated shape, and is lined by a membrane which itself forms the *membranous labyrinth*. In addition to the *cochlea*, which is that part of the labyrinth concerned with the perception of sound, certain histologically closely related structures here compose the *non-acoustic labyrinth*. In the frog, the *utricle* is thought to be concerned with the perception of static position and with movement relative to gravity (*e.g.* Dusser de Barenne, 1934), and the *saccule* with the perception of vibration of low frequency (Ashcroft and Hallpike, 1934). The three semicircular canals have been shown to be concerned with the perception of rotational movement of the head (*e.g.* Ross, 1936). Their functions are probably similar in man, and with these we need be concerned no longer here. The *acoustic labyrinth*, or cochlea, consists of a tube coiled into the form of a conical helix (Fig. 13), and in section consequently

, [1] Author's unpublished work, in collaboration with Dr C. S. Hallpike.

reveals a number of lumina in the containing bone. Each turn of the cochlea is known as a "whorl"; the human ear contains three such whorls, to which the names basal, medial and apical are conveniently given. The diameter of the cochlear tube decreases steadily from base to apex.

The lumen of the cochlear tube is divided into three compartments by two membranes, the *basilar membrane*, containing on its upper surface the *Organ of Corti*, and the thin *membrane of Reissner*. The three compartments thus produced are known as the *scalae tympani, media*, and *vestibuli* respectively (Fig. 15). On its interior side the basilar membrane is supported by the *osseous spiral lamina*, a thin shelf of bone running from base to apex of the cochlea. The external edge is similarly supported by the *spiral ligament*, a structure of considerable strength, and of greater size at the base of the cochlea than at the apex.

The organ of Corti comprises the rigid *arch of Corti*, supporting the inner row of *hair cells*, which is one cell thick, and the more numerous outer rows. These hair cells are thought to be the seat of nervous excitation. From them, nerve fibres run to the *spiral ganglion*, whence the exiting fibres unite in the *modiolus* to form the acoustic branch of the VIII nerve. From the upper surface of the hair cells of the organ of Corti a number of hairlets project. Over these lies the *tectorial membrane*, which is supported along its internal edge by a fibrous projection from the spiral lamina. Whether the upper ends of the hairlets are embedded in the substance of the tectorial membrane, or the latter merely touches these, has been a matter of controversy for many years. With improvements in histological technique, however, it has recently become increasingly evident that, at least in adult animals, the hairlets are actually embedded therein, and that the histological picture showing the tectorial membrane separated from them, found in the majority of text-books of anatomy, is to be ascribed to damage done to this exceedingly delicate structure during the preparation of the sections.

The scala vestibuli is separated from the air-filled middle ear by the oval window, to which the foot-plate of the stapes is attached, as has already been noted. At the apex of the cochlea the scala vestibuli is continuous with the lower scala tympani.

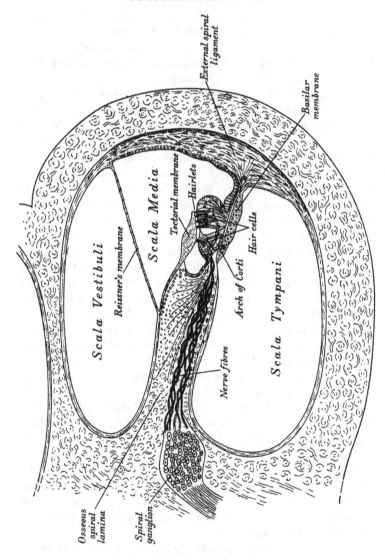

Fig. 15. Diagrammatic view of a section of a single turn of the human cochlea.

through a minute communication known as the *helicotrema*. Both these scalae are filled with the fluid *perilymph*, and its escape into the middle ear cavity from the scala tympani is prevented by the membrane of the *round window*. The latter acts as a relief membrane, releasing the increased pressure of the fluids caused by an inward movement of the oval window by bulging outwards. The movements of the two membranes, both of which may be seen from the middle ear cavity, are consequently 180° out of phase, an inward movement of one causing an outward movement of the other, and *vice versa*.

The perilymph of the inner ear is held to be similar to cerebrospinal fluid, the perilymph spaces being continuous with the subarachnoid space of the posterior fossa through the minute *cochlear aqueduct*. The scala media and the rest of the labyrinth, which are altogether separated from the perilymph system by the basilar membrane and Reissner's membrane, contain a second fluid, the *endolymph*, which is thought to be secreted within the endolymph system itself.

It is now necessary that some consideration be devoted to the mechanics of the cochlea. To facilitate this, it is important to recollect that the membranous labyrinth is contained in the rigid bony labyrinthine cavity, and that the fluids filling this are, for our present purposes, incompressible. Because of this, it will be seen that any pressure applied to these by an inward movement of the stapes foot-plate must be transmitted to the round window relief membrane, as has been noted above. This transfer of fluid might take place through the helicotrema connecting the two relevant scalae. The minute diameter of this, however, prevents such an event for all pressure changes save the slowest. The vibrations of the ossicles caused by the rapid changes of air pressure yielding an auditory sensation (the human auditory spectrum extends from some 20 to 20,000 c.p.s.) impart to the fluids variations of pressure very greatly too rapid for this channel of intercommunication to be of significance. A pressure change in the perilymph of the scala vestibuli can be released (by the round window membrane), therefore, only by distorting the elastic membrane of Reissner, thence through the endolymph, and so *via* the basilar membrane, itself a yielding structure, to the scala tympani. It is evident that the transfer of pressure changes across these membranes must involve that they themselves move. An increase in pressure of the perilymph of the scala vestibuli will depress the membrane of Reissner, thus increasing the endolymph pressure. This in turn will depress the basilar membrane, thus transferring the pressure change to the scala tympani, as has been said. Now it will be seen from Fig. 15 that the osseous spiral lamina will most probably hold the *internal* edge of the basilar membrane approximately still. In consequence of this, it has been argued (Wrightson, 1918) that the effect of changes of pressure imparted to the cochlear fluids by the stapes foot-plate will be to cause the organ of Corti to execute a rocking motion about its inward edge. On this theory the tectorial membrane is thought to be little affected by the changes of pressure involved, and will thus remain still relative to the rocking hair cells. By a downward movement of the basilar membrane the hairlets on these, which as we have noted above are probably embedded in the substance of the tectorial membrane, are thus bent backwards and pulled. Wrightson's theory is that this movement is responsible for the initiation of nerve impulses in the fibres leading from the hair cells, thus eventually yielding a sensation of sound.

Other views have been held as to the precise nature of the mechanism of stimulation of the hair cells, however. Thus, Helmholtz (1912) believed that the hairlets were not embedded in the tectorial membrane, but that stimulation occurred on the rising phase of the basilar membrane at that point when the hairlets were brought into contact with the tectorial membrane. Their stimulation was thought to take place in somewhat the same manner as that involving the production of a sensation of touch when a hair on the external surface of the body is lightly deflected. A third view has been urged by Hardesty (1908), who believes that the tectorial membrane itself is set into vibration, the basilar membrane remaining relatively stationary. On this hypothesis, stimulation occurs when the moving tectorial membrane touches the hairlets of Corti's organ. This view, however, has met with little acceptance. Hartridge (1934 a) proposes that, during the upward movement of the basilar membrane, the tectorial membrane is pushed up with it, and that during the downward movement the hairlets are pulled by the tectorial membrane in which they are embedded and which lags behind the basilar membrane. Recent investigations of the electrical phenomena of hearing, which will be discussed in more detail later (p. 95 *et seq.*), have shown that only one impulse in each nerve fibre is elicited for each cycle of the stimulating wave (Davis and Saul, 1932) and that stimulation occurs on the *rising* phase of the movements of the basilar membrane, corresponding to an outward movement of the stapes foot-plate (Davis, 1935). These discoveries permit the conflicting theories of stimulation of the VIII nerve to be differentiated with some certainty. Thus, we must abandon a majority of the earlier suggestions, since most of them demand that stimulation should occur on the phase corresponding to an inward movement of the oval window. Our present view as to the mechanism of stimulation may be sum-marised as follows. We know that one impulse is released per stimulus cycle, and that this occurs at a time corresponding to the development of negative pressure at the tympanic membrane, and thus to the rising phase of the basilar membrane. From mechanical considerations it becomes likely that the essential stimulation for the hair cells consists in the bending movement which must be imparted to their associated hairlets by the tectorial membrane, as the organ of Corti rocks upwards towards it. This, then, is the first stage in the development of an auditory sensation.

In the next chapter it will be our purpose to discuss the manner in which various theorists have attempted to account for the two most obvious characteristics of auditory experience, namely the perception of pitch and of loudness.

CHAPTER 6

The Perception of Pitch

Historically, theories of the operation of the ear have fallen into two main groups. It is not without interest that the first group of theories, which sought to ascribe to the peripheral organ those powers of frequency and intensity analysis which, in the second, were conveniently dismissed to the mysterious higher centres, has, in the past, had the more adherents. This fact has reacted favourably upon the amount of experimental activity in this field; for it will readily be appreciated that the theoretical ascription to a peripheral organ of powers of frequency and intensity analysis is verifiable more easily by experimental endeavour than is the opposing theory of central resolution. For this reason, the earlier part of this chapter, whilst it is intended to summarise all important experimental evidence on this topic, may appear, erroneously, to favour the theories of peripheral analysis at the expense of the remainder. This is solely to be ascribed to the preponderance of work on the former over work on the latter.

It is well known that the anatomical peculiarities of the cochlea have been made the basis of auditory theories by many workers in this field, from Contugno, in 1761, onwards. It must not be forgotten that, for the early theorists, as indeed for Helmholtz himself, the shape of the cochlea was virtually the only evidence in favour of a "resonance" hypothesis; almost without exception the remaining evidence has accumulated since their times. It is, therefore, a cause for considerable admiration that, by inspired deduction from this one salient fact, Helmholtz was able to enunciate a theory of hearing which, with some modification, still commands the majority of support from those interested in auditory function. His original exposition of the resonance hypothesis was simply that the arches of Corti, differentiated for size as they are, might be supposed to constitute a graduated series of resonating mechanisms. His attention, however, was shortly drawn to the fact that, in the bird, no structures comparable to Corti's arches exist, and for this and other reasons Helmholtz thereafter adopted the view that the basilar membrane itself constituted a differentially resonant structure such that high frequency stimuli yield vibrations localised at the basal end of the cochlea, and such that as the stimulus frequency is lowered the resonant region moves progressively towards the apex.[1] Helmholtz observed that the fibres of this membrane are connected together along their lengths by comparatively soft tissue, and he supposed therefore that the resonant structures, unlike those in a harp or piano to which instruments the mammalian cochlea is too often likened, are not sharply tuned, but are such that a note of any given frequency, whilst setting into vibration a considerable length of the membrane, nevertheless does so *maximally* at a point which is solely dependent upon the frequency of the stimulus. This matter has been the subject of a considerable amount of misunderstanding, for it has been felt that, if many fibres of the membrane are to be set into vibration at one time by one note, the assessment of its pitch on the basis of the identity of a *single* fibre could not well be possible. These and other inadequacies of the original theory were later met by Gray (1900), on what he has called "the principle of maximal stimulation", by supposing that the pitch of a note is recognised on the basis of the identity of those few fibres which are vibrating maximally. Gray made, also, other important contributions to the resonance theory. In 1900, he pointed out that most salient fact about the morphology of the cochlea, namely that the external spiral ligament is differentiated for bulk, and, therefore, apparently, for strength. He noted that, at the basal end, where the basilar membrane is least wide, the ligament is most large and dense, whilst at the apex, where the basilar fibres are longest, the ligament is smallest. It was therefore urged that the basilar fibres are differentiated not only for length but also for tension, a supposition which would appear very likely, having due regard to the limitations of morphology (Gray, 1900; Wilkinson and Gray, 1924).

[1] The best account in English of Helmholtz' resonance theory will be found in the translation of his original work (Helmholtz, 1912).

To Wilkinson and Gray, moreover, must go the credit for pointing out the implications of the fact that the basilar membrane is immersed in fluid; a feature of significance in that the additional loading thus applied to its fibres would enable those at the apex to resonate to frequencies much lower than they would be able to, were this loading absent. In this way it is possible to reconcile the facts that, whilst the lowest detectable sound is of frequency some 20 c.p.s., the longest fibre is under one-half a millimetre in length, much shorter than would be required in the absence of fluid loading.

At the opposite end of the frequency spectrum, the perception of a note of frequency 30,000 c.p.s. demands, in a resonant fibre about 0·16 mm. long, a stress of some 3·9 tons per sq. in. (Beatty, 1932, p. 47). As Beatty points out, this stress is not beyond the bounds of possibility, silk-worm gut having a breaking stress of 32 tons per sq. in., and spider thread one of 12 tons per sq. in. In summary it may be said, therefore, that the normal morphology of the cochlea is such as to provide peculiarly striking evidence in favour of a resonance hypothesis of function. It is essential, however, to note that this evidence is only of that somewhat indirect type which it is within the province of morphology to provide.

Of the remaining theories of peripheral analysis of frequency, that of Ewald is perhaps best known (Ewald, 1903; Ewald and Jäderholm, 1908). According to this, the pitch of a tone exciting the cochlea is perceived on the basis of the pattern of excitation of various parts of the basilar membrane. Thus, the basilar membrane is not excited to resonance over a limited area by a sinusoidal stimulus, as Helmholtz supposed, but vibrates in a pattern of some complexity over its whole length. Each tone produces, according to its frequency, however, its own unique pattern. For a combination of tones, the pattern is presumably still more complex. This theory has never met with general acceptance, either by physicists or physiologists, and has now been strongly contra-indicated by recent work on the nature of the response in the acoustic nerve. For this reason, it will be omitted from such further theoretical discussions as arise.

The second type of theory supposes that pitch perception

results from the presentation to the cortex of a clue as to the frequency of the stimulus, *qua se*. That is to say, on a modern formulation, the number of nervous impulses travelling up the auditory nerve in unit time will be identical with the frequency of the stimulus. Of recent years this theory has crystallised into the so-called "telephone" theory or frequency theory, whose most able exponent has been Boring (1926). Spatial analysis of the pitch of a tone is not here allowed for, for it is held that all tones will first excite the basal region of the basilar membrane, irrespective of frequency, the area of excitation growing steadily larger as the *intensity* of the tone is increased. Thus the apical hair cells will be excited only by tones of great intensity, the peripheral intensity clue being the length of basilar membrane vibrating. It is supposed that the wave form of a complex tone is in some manner transmitted to the higher centres of the brain as such, in a fashion difficult to envisage in terms of present day concepts as to the nature of the nervous message. This theory has consequently been subject to much criticism, especially from physiologists (*e.g.* Adrian, 1928). It is safe to say, however, that prior to 1930 both the telephone and the resonance theory were very widely held. In this year, the position was materially changed by certain discoveries in the experimental field (Wever and Bray, 1930*a*, *b*, *c*, and *d*), as a result of which combinations of the resonance theory and the frequency theory have been effected. Of these the best known example is the so-called resonance-volley theory, to a discussion of which we shall come later in the present chapter. Meanwhile, it will be fruitful to review the experimental evidence for and against the opposing classical theories of audition.

It has been known for many years that continued exposure to intense auditory stimulation produced certain pathological conditions within the cochlea. It is not unreasonable to suppose that, if pure tone exposure stimuli are employed, such cochlear lesions would be localised to those parts of the basilar membrane which, on the resonance theory, are expected to be set into maximal vibration by tones of the appropriate frequency. A considerable amount of experimental evidence on this point has become available, and this is, in the main, in favour of the resonance view. It is convenient to group these data according as they have arisen from studies of animal or human material.

In connection with animal studies, it is customary to cite the work of Yoshii (1909), (*vide, e.g.*, Beatty, 1932; Hartridge, 1934a); it is not uninteresting to note, therefore, that the earliest work on this subject appears to be due to Wittmaack (1907). The method used by these investigators was simple enough. After exposure to whistle, siren, or organ pipe generated tones, for varying lengths of time, the animal (usually a guinea-pig) was killed and a histological examination of the cochlea carried out, by the method of serial sections. Both investigators found destruction of the organ of Corti in all animals in which the exposure tone had been applied for a sufficient length of time, and it was found, moreover, that the regions of destruction occurred in order of frequency; that is, those due to high tones took place in the basal turn, whilst those due to progressively lower tones occurred at positions progressively nearer to the apex. The lesions were of considerable extent, usually covering one-quarter to one-half a turn of the cochlea. This last observation led Yoshii (*loc. cit.*) to suppose that the affected region was too broad to offer support to Helmholtz' theory, and Yoshii himself preferred, for that reason, the theory of Ewald (1903). It has been pointed out (*e.g.* by Hartridge, 1934a) that Yoshii's rejection of the Helmholtz theory was due to one of the most common misapprehensions as to the nature of Helmholtz' suppositions. It is certain that the size of the area of basilar membrane injured by any given exposure tone would be, in part, a function of the *intensity* of the tone. It is clear, therefore, that with the loud sounds used by Yoshii, it would be natural to expect a considerable area of lesion. Moreover, as Hartridge has urged (*loc. cit.* p. 937), Yoshii's use of whistles, an organ pipe, and a siren, all of which instruments are sources of somewhat impure tones, of variable frequency, renders the expectation of a large area of damage all the more reasonable.

The work of Wittmaack and of Marx is of substantially similar import. In summary, it may therefore be said that all three investigators failed to secure recognisable lesions in the apical turn, to low frequency stimuli. At least on occasions, however, higher tones yielded histologically detectable disturbances occurring in order of frequency. Later workers have failed to confirm these results (for summary, see Kemp, 1935).

Horton (1933, 1934) describes a series of experiments in which the auditory sensitivity of guinea-pigs was tested by the conditioned response method, after exposure for 110 hours to a note of frequency 1024 c.p.s., and intensity of 125 db. above the human threshold. He states that the hearing loss "was no more confined to one tone than to another" (Horton, 1934), a statement which his published audiograms fully substantiate. He tested his animals also at the eight octave tones from 64 to 8192 c.p.s., and it is apparent that they showed great variability of response both from frequency to frequency, and from test to test at the same frequency. Indeed, the only safe conclusion that may be drawn from his data is that an exposure tone of a given frequency may adversely affect the ear's sensitivity to tones of other frequencies, which need not necessarily approximate to that of the stimulus. Horton does not bring any histological evidence to bear upon his data in the papers referred to.

Wever, Bray and Horton (1934) and Davis, Derbyshire, Kemp, Lurie and Upton (1935) have brought, in addition to conditioned reflex tests and histological examinations, a further technique to bear upon the problem. In many of the animals used by these investigators (as before, guinea-pigs have been almost exclusively employed), a so-called "electrical audiogram" was obtained, as well as a normal audiogram obtained by the method of conditioned reflexes. The electrical audiogram consists of a curve relating the frequency of a pure tone to its intensity such that a just perceptible electrical response is obtained from the cochlea.[1] A fuller description of this technique is given by Lurie, Davis and Derbyshire (1934). It is sufficient to point out here that such a curve can provide evidence only as to the integrity of the peripheral organ, and will in no case indicate a loss of function more centrally mediated. In the opinion of Davis and his coworkers, the "audiograms" obtained by this method agree

[1] The electrical phenomena of the cochlea are discussed below, pp. 95–106.

"within experimental error, from 250 to 10,000 c.p.s. with the threshold of response by the method of conditioned reflexes" (Davis *et al. loc. cit.* 1935 *b*, p. 274).

The results obtained by these methods may be briefly summarised. Relatively widespread damage is done to the cochlear contents, in particular to the external hair cells and occasionally other parts of the organ of Corti, by exposure of the animals to protracted stimulation at high intensities. The thresholds of the electrical response are found to be higher than normal by an amount which corresponds roughly with the severity of the cochlear damage. In almost all cases, irrespective of the frequency of the stimulus, *functional* impairment of hearing, as tested by the conditioned response method, centres at approximately 1200 c.p.s. For the higher stimulating tones (2500 c.p.s.), the damage centred in the middle of the second cochlear whorl, and was correlated with impaired *electrical* sensitivity at various frequencies centring about the stimulus frequency. Within limits, therefore, the resonance theory is indicated, at any rate for the higher tones, and the normal audiogram is shown to be an untrustworthy indicator of peripheral auditory function, under these pathological conditions.[1] Facing these somewhat inconclusive results, it is satisfactory to be able to turn to the human pathological evidence.

References to the pathology of high-tone deafness, and to the importance of the evidence obtained therefrom for the resonance theory of hearing are frequently met with in physiological textbooks. The evidence is, in brief, that boiler-makers and others

---

[1] In a personal conversation with the author, Dr Davis ascribed these somewhat disappointing results in part to what he called a "mechanical short-circuiting effect". Thus the sensitivity for *all* stimuli below a certain frequency was reduced owing to the fact that the basilar membrane had, by the exposure, become much weakened at a point in about the middle of its length, corresponding to the point of maximal sensitivity of the ear. In consequence fluid transmitted pressure changes tended to pass through this region, rather than set up localised vibration at any point nearer to the helicotrema, as would be expected for the lower tones. This supposition undoubtedly accounts for the generally lowered sensitivity for low frequency stimuli, exhibited by the majority of animals. It fails, however, to account for the singular centring of the functional impairment, in all animals, at approximately 1200 c.p.s., which it is necessary to ascribe to a "central" deafness.

whose normal *habitat* is one in which loud, high frequency noises
are usually present, suffer, with advancing age, an increasing
deafness to high tones. This loss of sensitivity may be correlated
with histological changes in the organ of Corti, and occasionally
elsewhere, in the basal turn of the cochlea. Unfortunately, this
evidence has not been sufficiently precise to contra-indicate the
frequency theories of audition, for it is clear that, as these
changes take place nearest to the oval window, they might also
be accounted for on the latter theories, as taking place at that
part of the cochlea at which the amplitude of vibration would be
greatest. Recently, however, Crowe, Guild, and Polvogt have
made a study of such precision, and of so much material, that
they have been enabled to correlate changes in functional
sensitivity, measured audiometrically, very precisely with
changes in histological appearance; and, moreover, to do so in
such a way as to be able to ascribe to various parts of the human
basilar membrane in the first cochlear whorl the frequencies to
which, under normal conditions, they may be supposed to
resonate (Guild, Crowe *et al.* 1931; Guild, 1932; and especially
Crowe, Guild and Polvogt, 1934). This has been possible only for
the higher frequencies of the auditory spectrum, and a similar
attempt to correlate low tone losses with apical turn lesions has
not been so successful (Guild, 1935). It will be shown later,
however, that evidence on the latter point has arisen in another
fashion, and it is, therefore, of the greatest importance that these
workers, by so good a technique, have been able to provide exactly
such evidence as was required to establish that, for high tones at
least, the mammalian cochlea contains a differentially responsive
structure.

Employing a somewhat different method, Held and Kleinknecht (1927) have also succeeded in producing evidence in
favour of the resonance view, with regard to the perception of
high tones. These workers succeeded in boring a very small hole
in the basal turn of the guinea-pig's cochlea, employing drills of
an average diameter of 0·1 mm. The drill was permitted to
penetrate only a short way into the cochlear contents, and thus,
in the successful cases, produced localised damage only to the
external spiral ligament, as was demonstrated upon subsequent

histological examination. As a result of this operation, the animals, whose hearing was tested by making use of the pinna reflex, were deaf to a narrow band of tones of high frequency. Held and Kleinknecht believe that their operation had only released the tension on the basilar membrane over a small area, and they concluded, therefore, that not only was the membrane tuned, according to customary expectation, but also that the tension exerted by the spiral ligament was fundamental in maintaining this tuning.

This work has been criticised in later discussions, mainly on the grounds that the use of the pinna reflex is unsatisfactory as an indicator of auditory performance. It is of importance, therefore, that the experiments have been repeated in substantially the same fashion, using, however, a technique at which this criticism cannot be levied, by Stevens, Davis and Lurie (1935). In these experiments, so called "electrical audiograms" (*vide supra*) were taken in guinea-pigs, both before and after producing local injury to the organ of Corti, by the use of fine drills. The experiments were not restricted to the basal end of the cochlea, as were Held and Kleinknecht's, but injuries were produced in various parts of each cochlear whorl. As a result, these workers found that, over a restricted band of frequencies, no response, or at best only a much reduced one, could be elicited from the injured cochlea. The position of the band of insensitivity on the frequency scale was dependent upon the spatial position of the injured area, and the authors of the paper referred to conclude (elsewhere) that "High tones are localised near the basal end of the cochlea; 2000 c.p.s. in the middle; and low tones, to which we directed particular attention, are bunched together quite closely toward the helicotrema" (Davis, Lurie and Stevens, 1935, p. 777).

Many experimenters have endeavoured to manufacture working models of the cochlea. Helmholtz, M'Kendrick, Ewald, and Wilkinson are among these; data derived from the operation of the models of the latter two investigators have been the subject of considerable discussion, and short consideration may therefore be given to them.

Ewald's model (Luciani, 1917, p. 241) consisted of a brass box, in which a diagonal partition was fitted. A long narrow slit was cut in this partition, and covered by a membrane of thin rubber sheeting. Two further rubber membranes were arranged upon the external surfaces of the box, and the whole was filled with fluid. A glass window was arranged so that the behaviour of the internal membrane in response to vibratory activity of one of the external membranes could be observed by means of a microscope. From this assembly, Ewald claimed to have obtained evidence of the most indisputable character pointing to his "sound-pattern" theory of hearing.

Wilkinson's model differed in detail from Ewald's. In this, a metal box was divided as before, but by a partition composed of a frame on which were wound a number of turns of wire, representing the basilar membrane fibres; like the latter, these were differentiated for length and tension. A thin membrane was pasted across them, to resemble their transverse connections. As before, two membranes were arranged on the external surface of the box, and vibrations were transferred to the contained fluid through one of these. Observation of the partition representing the basilar membrane was permitted by the heaping up of small particles of enamel. Wilkinson and Gray (1924) believed that this model provided evidence for the resonance hypothesis, but Hartridge (1934 a) claims that, on the basis of four sets of photographs in his possession, the apparatus more often showed multiple responses, that is responses at more than one part of the basilar membrane, than single ones. He is therefore of the opinion (loc. cit. p. 939) that this model showed more evidence for Ewald's than for Helmholtz' theory.

It is important to note that, in more than one way, these models cannot provide a true representation of the cochlea. Physically, the constants governing the properties of the majority of materials employed are vastly different from those of the materials of the ear. It has, moreover, been pointed out (Hartridge, loc. cit. p. 939) that the vast disparity in size between the cochlea and the models precludes the possibility of any very direct evidence being obtained. The fact that, as Hartridge says, "Not a single one of any of these models has failed to substantiate the theory of its designer..." (loc. cit. p. 938) is sufficient to warrant the exclusion of the evidence of models, in considering cochlear function.

Hartridge's own experiments on the effect of a phase change of one-half wave-length on the human ear may now be discussed. In his earlier work (Hartridge, 1921) he used a modified de la Tour siren, so arranged that, by rotation of the wind chest through 10° about the same axis as that of the siren disk, a phase change of 180° was introduced into the otherwise continuous musical tone emitted by the instrument. When this occurred, a sudden change in the intensity of the note was produced, such that it appeared to fall to zero and return to normal strength in a fraction of a second. As Hartridge deduces, this effect, character-istic as it is of the behaviour of a resonant vibrating system, makes it likely that the cochlea does act as a resonator, as is demanded by the Helmholtz theory. In a later repetition of the experiment (Hartridge, 1934b) substantially similar results were found, using quite different apparatus. A brass disk of diameter 1 ft. is arranged to rotate truly on a steel shaft. An inner row of ninety-six exactly equidistant slots is milled through the disk. An outer row of ninety-six slots is then milled in such a way that the distance between the slot first and that last milled is only one-half the distance between any other pair of slots. If a stream of air is directed through the inner row, with the disk rapidly rotating, a continuous note is produced, in the typical siren manner. If the air jet is directed on to the outer row, however, the note is interrupted once in each revolution of the disk, when the phase change of 180° occurs, due to the two slots at half distance. The theoretical implications of this experiment are, of course, exactly similar to those deduced from the first apparatus.

Experiments in which the relative phase of the components of an impure tone is changed have been made by Cosens and Hartridge (1922) and by van der Pol (1929). It is clear that, when, say, two pure tones are combined in different phases, the wave form of the resultant vibration must vary very greatly. An excellent graphical illustration of this is given by Beatty (1932, p. 61). Cosens and Hartridge arranged two forks, of frequency 142 and 426 c.p.s., in such a manner that the vibrations of one might be stopped and started in random fashion. The phase relationships of the two resultant tones might be expected also to vary, yet no change whatever in the quality of the note could be detected. It is clear that, on the resonance theory, no mechanism is provided whereby the ear could detect a change of phase, whereas if all frequencies are transmitted as such to the higher centres, the supposition is that a phase change would be noticeable. Van der Pol has devised a system, using thermionic valves, whereby the phase of the high-frequency components of a complex tone may be retarded in varying amount relative to that of the low components. He has been unable to produce any alteration in the quality of music transmitted through this equipment, and an amplifier and loudspeaker. The observation cannot be escaped that the ear must be quite insensitive to such alterations of phase, and, in so far as this is an argument at all, it must be thought to indicate the resonance theory of hearing. It must not be thought, however, that the frequency theories are strongly contra-indicated. There is no reason to suppose that, even if it were presented with indications of frequency as such, the auditory cortex must necessarily be sensitive to monaural phase changes; although the evidence from experiments on sound localisation is such as to indicate some sensitivity to binaural phase relationships (Stevens and Newman, 1936), or time intervals (Banister, 1926 a, 1926 b, 1926 c).

Important evidence strongly in favour of the resonance view has recently been derived from the study of the relation of pitch to intensity. It has been known for some years that the perceived pitch of a sound is not only dependent upon its frequency. Thus, Zurmühl (1930) showed that, for stimuli from 200 to 3000 c.p.s. an increase in intensity produced a slight lowering of pitch. At his highest frequencies, Zurmühl showed that the effect was very slight, and indeed was often reversed in direction. More recently, Stevens (1935) has published the results of a very complete investigation on one human subject, and has shown that, for sinusoidal stimuli of any frequency between 150 and approximately 2000 c.p.s., an increase in intensity produces a lowering of pitch, results in substantial agreement with Zurmühl's. For stimuli above 2000 c.p.s., moreover, the opposite effect is apparent. Here, an increase in intensity results in increase in pitch; that is to say the note becomes somewhat sharp. It may be seen from Stevens' results that there is no sudden change over from an effect in one direction to an effect in the other, as the frequency of the stimulus is raised, but that over the middle frequency range, that is to say from 1000 to 3000 c.p.s., there is an area over which intensity has extremely little effect on pitch. Outside this area, the effect increases steadily as one proceeds up or down the frequency scale, becoming positive in the former and negative in the latter case, as has been said.

These results are of great importance to the theory of frequency perception. It has been shown by various investigators that the frequency of the impulses in the VIII nerve of the guinea-pig and the cat[1] is quite unaffected by changes in the intensity of the stimulus. On the telephone hypothesis, pitch is judged on the basis of this frequency, however, and the fact that it is in no way dependent upon stimulus intensity argues against this formulation strongly. Troland (1930) has, indeed, suggested that there might be some sharpening of pitch with an increase in intensity, due to the increasing introduction of subjective harmonics, but even this hypothesis fails to account for the fact that perceived pitch can move in either direction, as Stevens (*loc. cit.*) points out. This evidence is, therefore, strongly in favour of the opposing resonance theory, for it is most clearly shown that perceived pitch and impulse frequency are not related. Stevens shows that the point of maximal stimulation of the basilar membrane is shifted towards its nearer end as the stimulus intensity rises, assuming Fletcher's (1930) calculations as to the position of maximal resonance for various frequencies to be correct.

The literature on the subject of auditory theory contains many objections to the resonance hypothesis. The majority of these have arisen from misconceptions as to the nature of Helmholtz' theory, or as a result of evidence deduced from experiments scarcely relevant to the subject. For these reasons, a writer attempting to summarise the evidence for and against the extant theories must find himself in the position of discovering much

---

[1] *Vide infra*, pp. 95–106.

objection to the best known of these, but very little evidence of importance in contra-indicating such a concept. A most prominent objector, in the early literature at least, is Keith. In the appendix to Wrightson's *An Inquiry into the Analytical Mechanism of the Internal Ear* (Wrightson, 1918), this writer makes, *inter alia*, the following objections:

(*a*) That the whole auditory spectrum could not be covered by a series of resonators varying in length by so small an amount as do the basilar membrane fibres. Among others, however, Beatty (1932, p. 46) has shown that no impossible tensions are involved if this supposition is made (*vide supra*, p. 82). By a slightly different process of analysis, Hartridge (1922) arrives at figures even more within the sphere of likelihood.

(*b*) That there exist "no anatomical structures which can serve as resonators in the cochlea", as "the basilar fibres are not free strings, they are clothed on their lower surface by a layer of loose cells and fibres, and their upper surface is clothed by a stratum of cells" (Keith in Wrightson, 1918, p. 187). The answer to this objection is given simply by Beatty (1932, p. 44) who points out that the superstructures referred to must act like the helix of copper wire so familiar on the bass strings of a piano, where it lowers the resonant frequency of the string without any additional length being required.

In the book referred to, Wrightson makes the familiar objection to the resonance theory that, as the fibres of the basilar membrane are connected together along their lengths, the sharpness of resonance would be so negligible as to preclude any tonal perception on the basis of the identity of the vibrating fibre. This objection has already been mentioned, and it is clear that an adequate answer to it is provided by Wilkinson and Gray's "principle of maximal stimulation", whereby it is supposed that pitch is recognised on the basis of the identity of that fibre vibrating *maximally*.

Hartridge (1934*a*) quotes Kolmer as stating that a single fibre of the auditory nerve may branch to supply more than one separate hair cell. He points out, also, that Held concludes that a single nerve fibre may be distributed over a quarter of a turn of the cochlea. He notes that this may indeed be the case for the extreme apex and base of the cochlea, for pitch discrimination at the extremes of the spectrum is anyhow poor (Knudsen, 1923; Shower and Biddulph, 1931). On such a basis, moreover, it is possible to meet this important objection to the resonance hypothesis, for there is some reason to suppose a substantially augmented innervation for the middle cochlear range, as may be seen from the relevant literature on the anatomy of the VIII nerve, and its connections.

The data of Guild, Crowe, Bunch and Polvogt (1931) and of Lorente de Nó (1933) make it apparent that at least two classes of nerve fibres may be recognised in the organ of Corti. These differ to some extent in their distribution as well as in their mode of connection. The most numerous type is the "radial", the fibres of which pass straight to Corti's organ, and make connection with the internal hair cells. Each fibre is connected with only a few cells, and each cell is innervated only by the branches of a few fibres. The second group includes "spiral" fibres of several types, which are considerably less numerous. These run for varying distances up to a third of a turn of the cochlea, and each fibre establishes connection with many external hair cells. It is clear, therefore, that only the distribution of the radial fibres provides evidence which would support the view that the basilar membrane is projected spatially in the midbrain, though for the existence of such a projection Lorente de Nó (*loc. cit.*) has what would appear to be good evidence.

Recently, moreover, Ades, Mettler and Culler (1938) have apparently demonstrated a point-to-point connection between the basilar membrane and the medial geniculate body, very much nearer the cortex. And, lastly, Poliak (1932) and Walker (1937) have additionally demonstrated that a projection of the latter upon the auditory cortex (superior temporal convolution) shows a further point-to-point correlation. The conclusion cannot be escaped that the existence of these apparently discrete paths from the basilar membrane to the cortex provides excellent evidence for a theory of peripheral frequency analysis.

A further objection to a resonance hypothesis is advanced by Troland (1929), among others. Considering the evidence obtained from patients with so called "islands of deafness",

Troland points out that such a person can correctly perceive the pitch of a note to which he is normally deaf, if the note is made loud enough. This should not be the case, for it is clear that the increase of stimulation intensity should lead to vibration of basilar membrane areas adjacent to that normally associated with the perception of the frequency in question; and, as these latter are functional, they will provide the "point of maximal stimulation" clue required. In brief, on the resonance theory, a person with such a deafness as that under discussion should perceive, not the frequency to which he is deaf, but an adjacent one which he is normally able to perceive. Hartridge (1934 a) argues that this phenomenon may be explained if one assumes that Troland is mistaken in his concept of the pathology of this type of deafness. Hartridge assumes that, for intense stimuli, the increased amplitude of vibration of the affected part of the basilar membrane is sufficient to stimulate that part. It is clear both from general physical considerations and from the results of recent experimental work (Davis, 1935, p. 214), that the vibration of the basilar membrane is very heavily damped, and thus that a large part of the membrane will vibrate in response to any one intense pure tone stimulus; it is therefore difficult to see why an unaffected part of the organ of Corti should not have given an erroneous frequency clue before the pathological area is stimulated. It is difficult to feel with Hartridge that "Troland's criticism therefore falls to the ground" (loc. cit. p. 938), but it may here be pointed out that as yet there is quite insufficient evidence both on the question of pitch perception in patients suffering from this somewhat rare disease, and on the pathology of the disease in general.

It will now be our purpose to examine the data bearing upon the question of frequency and loudness perception, which have arisen from the electrophysiological examination of auditory activity.

In 1930 Wever and Bray (1930 a, b, c and d) announced the discovery of what they believed to be action potentials in the VIII nerve of the cat, and the further discovery that these potentials accurately reproduced the frequency of the incoming stimulus. On this evidence it was generally thought that

Helmholtz' theory had been strongly contra-indicated. The later discovery by Adrian (1931 *a*, *b*) that the greater part of the potential changes did not arise in the VIII nerve, but were due to what he called the microphonic action of the cochlea, tended to remove the objection. Davis and Saul (1931, 1932), however, later gave what appears to be conclusive proof of the participation of the intracranial fibres of the VIII nerve in the production of the Wever and Bray effect; they showed, too, that at least up to a frequency of 2000 c.p.s. the impulses in the auditory nerve were synchronised to successive waves of the incoming stimulus. This apparent proof of Wever and Bray's contentions seemed so conclusive as to lead Adrian (1932, p. 40) to say, "It is, I think, an open question whether there will be much left of the resonance hypothesis of the cochlea when Wever and Bray have finished their investigations." It must be our next purpose therefore to examine Davis and Saul's evidence on the question of synchronised action potentials.

In their original papers, Davis and Saul (1931; Saul and Davis, 1932) stated that they had been able to detect, in the midbrain of the cat, action potentials whose frequency, up to the middle range of the auditory spectrum, was the same as that of the sound stimulus. The evidence they brought forward was convincing in indicating that these potentials were due to the passage of nerve impulses; thus they were deleted reversibly by narcotics and by chloroform anaesthesia, irreversibly by death of the animal; they were quite strictly localised to the auditory tracts, and could not be detected in other parts of the midbrain, as would be the case if they were due to mere electrical spread from the cochlea. They pointed out also that, although the cochlear response is well maintained up to several thousand cycles per second, the *synchronised* response in the midbrain disappears when the stimulus frequency rises much above 1000 or 2000 c.p.s. At frequencies above this, a response is obtained, but this appears to be quite asynchronous, and is consequently more difficult to detect. In the later paper of the two under discussion, Saul and Davis (1932) differentiated clearly between the cochlear response and the true nerve response. All succeeding work has gone to prove the validity of this differentiation.

The detection of nervous impulses at so high a frequency as 2000 c.p.s. naturally led to considerable discussion. There is every reason to suppose that the absolute refractory period of most mammalian medullated nerve lies between 0·6 and 1 ms.[1] (*vide, e.g.,* Gasser and Erlanger, 1925; in this paper the absolute refractory period of the phrenic nerve of the dog is given as 0·6 ms.). The reproduction of a frequency of 2000 cycles involves the supposition that the refractory period of the VIII nerve is less than this latter value. As Adrian pointed out (1932, p. 39), many of the fibres of the VIII nerve are of great diameter (up to 18 μ.),[2] and great size is, in other nerves, an indication of shortness of refractory period. It appears unlikely, however, than on analogy with other mammalian nerve fibres whose refractory period has been measured, a value as short as 0·4–0·5 ms. could possibly be ascribed to the fibres of the nerve in question. Wever and Bray (1930*d*; Wever, 1933) have, however, suggested a possible mechanism whereby the phenomenon can be accounted for. This hypothesis they called the Volley Theory, and such a theoretical mechanism must now be described.

In brief, it is pointed out that, as the VIII nerve is a trunk of considerable size and contains many fibres, it is possible to suppose that the responses in individual fibres are "staggered"; that is to say, whilst one fibre is in its refractory phase, having just conducted an impulse, the next is itself conducting; by the supposition of a regular alternation of response in this fashion, it is possible to account, even in a trunk of relatively few fibres, for the reproduction of the highest frequency so far found. It must here be pointed out that this theory was originally put forward to account for the Wever and Bray effect itself, then thought to be of nervous origin. Frequencies at least as high as 8000 c.p.s. had, therefore, to be accounted for. With the later discoveries that the true nervous response was apparently restricted to frequencies below 2500 c.p.s., the theory provided an even more adequate description of the experimental data. In almost unmodified form, indeed, this theory is used by Davis

---

[1] ms. = millisecond = 10^{-3} second.
[2] μ. = 10^{-3} millimetre.

(1934) to account for the phenomena of equilibration and alternation in the nervous response, to which consideration will shortly be given.

It may well be considered that such a theory as that above is purely speculative; the alternative theory, supposing for the VIII nerve a somewhat shorter refractory period than has hitherto been found in other mammalian nerves, may appear to be no more speculative, and, by reason of its greater simplicity, no less attractive. Davis and his co-workers have, however, succeeded in adducing two extremely important pieces of evidence which argue against this view, and in favour of a volley mechanism (Davis, Forbes and Derbyshire, 1933; Davis, 1935; Derbyshire and Davis, 1935). In order to discuss this evidence, it is necessary to consider the order of events in the auditory nervous response, after the onset of a sustained stimulus. After a brief initial transient, which has been referred to as the "on-effect", the impulses become synchronised to the frequency of the stimulating wave, provided, as has been pointed out above, that this frequency is not above some 2000 c.p.s. For the ensuing tenth of a second, or less, there is a rapid diminution in the amplitude of discharge, which finally settles down asymptotically to a steady state.[1] Both the size of the initial wave after the "on-effect", and the amount of the shrinkage, or "equilibration", are closely related to the frequency of the stimulus. The nature of this relation is illustrated in the graphs shown in Fig. 16 (a) taken from the paper referred to (Derbyshire and Davis, 1935). From this it may be seen that below about 1000 c.p.s., the response size is approximately constant, but that at this frequency it commences to fall steeply. A similar though less marked fall occurs at a frequency of about 2000 c.p.s. An equally indicative relation obtains between the amount of equilibration and frequency (Fig. 16 (b)). From these data, Davis and his co-workers make the following deductions. For all frequencies below that at which the first drop occurs (in the graph shown, 1000 c.p.s.), every nerve fibre can follow the frequency of the stimulating sound. During continued stimulation of these fibres at a frequency near the limit imposed upon them by their

[1] This is true only for moderate stimulus intensities: *vide infra*, p. 117.

refractory period however, this latter value tends to increase. As Davis points out (1935), this is quite a well-attested fact for other mammalian nerve fibres, and can therefore be assumed to be the case for those of the VIII nerve. It is clear, then, that if the refractory period tends to lengthen, all fibres cannot continue to respond at the initial high frequency, and therefore

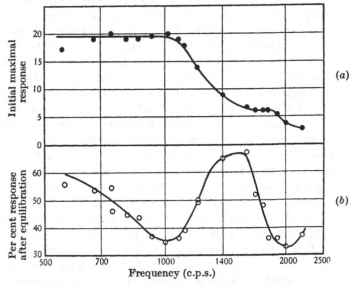

Fig. 16. The upper curve shows the relation between frequency of stimulus and size of the initial response in the auditory nerve of the *cat*. The measurement was made immediately after the "on-effect". The lower curve shows the percentage of the initial response remaining after 2 sec. stimulation, and thus gives a measure of the amount of equilibration (after Derbyshire and Davis).

they commence, one by one, to respond to each *alternate* wave. This phenomenon is termed "alternation", and Davis points out that it adequately accounts for the rapid initial equilibration previously described.

For tones of higher pitch than about 1000 c.p.s., no fibre can respond to each stimulus wave, but some respond to every other wave. The amplitude of the electrical response is, therefore, only about one-half that produced by a lower frequency stimulus. At a frequency approximately double that at which the initial alternation drop occurs, that is about 2000 c.p.s. in the case considered, a still further decrease in response is manifested; this is thought to correspond to the point at which the fibres are forced from their previous 1 : 2 relation to a 1 : 3 relation.

It is obvious that the exact frequency at which this alternation occurs will depend very materially upon such matters as temperature and general condition of the nerve; it is, for example, invariably accepted that at lowered temperatures the refractory period of nerve tends to rise. In general, however, Davis and his associates have found that the first alternation occurs between 800 and 1000 c.p.s., and the second at double these frequencies; they deduce therefore, that the refractory period of the fibres of the cat's VIII nerve lies between 1 and 1·2 ms., values similar to those for other mammalian nerve fibres of comparable anatomical characteristics.

This evidence, which will be discussed again in the next chapter, would appear to put the volley theory upon a sound footing; such a theory provides an adequate explanation of the phenomenon of frequency reproduction in the VIII nerve response. It remains, however, to discuss the implications of this finding upon existing theories of the perception of pitch.

At first sight, the fact that the frequency of the incoming sound stimulus is reproduced in the form of synchronised nerve impulses must be thought to place the frequency theories of perception upon a sounder basis than they have hitherto occupied. Indeed, the only limitation upon such a supposition must be thought to lie in the evidence that only frequencies up to approximately 2000 c.p.s. are so reproduced. For this latter reason, another mechanism than that inherent in the frequency theory must be supposed for the perception of frequencies above 2000 c.p.s.; that is to say, for frequencies above this, a resonant mechanism must almost certainly be called into play.

On the other hand, it may well be argued that, if it is necessary to postulate the resonance mechanism for frequency perception in the upper register, does it not unnecessarily complicate auditory theory to suppose a totally different mechanism for the perception of lower frequencies? The finding that some frequencies are reproduced in the auditory nervous system may well reflect nothing more than the intermittent character of the stimulus acting on the sensory cells of the organ of Corti. Hartridge (1934 b) has pointed out, moreover, that owing to the fact that individual fibres of the VIII nerve and tracts are of different sizes, the conduction rates in different fibres must also vary. Thus, whilst one may find synchronised responses when recording from a nervous region near to the ear (for example, the VIII nerve), when recording from a region nearer to, or in, the cortex, no such synchrony would, it is supposed, be found; for the times of arrival of individual impulses corresponding to one and the same stimulus wave would be quite different, owing to their different rates of propagation. This supposition has been confirmed by Davis (1935), who has failed to detect any trace of synchronous activity in the cortex of the cat, for any stimulus frequencies higher than 20 c.p.s. As he points out, such a frequency sounds to the human ear more like a flutter than a sustained musical tone. The suggestion is obvious, therefore, that the presence of synchronous activity in the cortex corresponds to a buzzing or fluttering sensation, and not to the sensation produced by a musical tone. Such evidence as this argues against a telephone theory, with regard to the discrimination of frequency, except for the lowest audible tones. It will be shown in the next chapter, however, that the experimental evidence is somewhat more strongly in favour of a mechanism such as that supposed in this theory, for the perception of loudness. It may well be, moreover, that for stimuli of extremely low frequency the synchronous character of the VIII nerve discharge is of importance for other reasons. The resonance-volley hypothesis has indeed been employed to explain certain phenomena noted in the human subject for stimuli below 60 c.p.s. (Wever and Bray, 1936).

The electrical activity of the cochlea has been employed as an indicator in an attempt to demonstrate the existence of a resonant structure in the living animal. It has already been pointed out that the cochlear component of the electrical activity detected at the round window arises as a result of the vibration of some internal structure or structures of the cochlea.

Now the resonance theory postulates that the basilar membrane is tuned for frequency in the manner earlier discussed in some detail. Consequently, the point at which its amplitude of vibration is maximal will vary along its length, if the frequency of the stimulus is changed. It is extremely likely, in addition, that any other vibrating structures in the cochlea will also vibrate maximally at points opposite the maximally vibrating areas of this membrane, the fluid resistance to their motion there being minimal. If this is the case, it must be supposed that the point at which the potentials constituting the cochlear electrical effect arise will vary along the length of the cochlea, their point of greatest magnitude being near the base for high frequencies and near the apex for low. Should it be possible, therefore, to demonstrate that the potentials obtained from an apical electrode and from a basal electrode (with reference to a common return placed elsewhere) show such a differentiation as this, the presence of resonant mechanisms within the cochlea would be strongly indicated.

It will be appreciated that low impedance electrical contact with the basal end of the cochlea can readily be secured by an electrode on the rim of the exposed round window. Contact with the apex presents, in a majority of animals, somewhat greater difficulty, owing to the dense and avascular character of the containing temporal bone, whose characteristics render it a good electrical insulator. In order to establish contact here, therefore, it is necessary in some manner to reduce the thickness of this bone, yet without disturbing the cochlear contents or permitting the escape of endolymph or perilymph fluids. By drilling a small hole with a dental burr, until such a time as its floor was very thin, and filling the resultant cavity with a small mercury bead, these necessary conditions were fulfilled, in the cat, by Hallpike and Rawdon-Smith (1934 b, c). From preparations such as this results illustrated by the graph of Fig. 17 were obtained, showing quite conclusively a crossing over of the two response curves. Very similar results were later secured by Stevens and Davis (Stevens, Davis and Lurie, 1935; Davis, 1935), using a guinea-pig preparation. This animal possesses the great advantages that the cochlea projects into the middle ear cavity, and that the bony covering at the apex is so thin that good electrical contact may here be established without resource to the operation of drilling.

Though apparently so indicative, the results of these experiments should be accepted with some caution. Thus, the differential effect noted appears somewhat sensitive to the precise

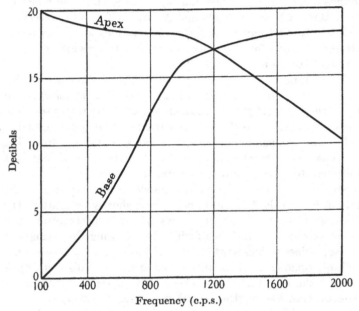

Fig. 17. Graphs showing the relative responses obtained from basal and apical electrodes on the cochlea of the *cat*, for various stimulating frequencies. Note that the curves intersect at approximately 1200 c.p.s.

placing of the recording electrodes, and in the majority of successful cases it would seem that the apical electrode is in most intimate contact with the scala vestibuli whilst the basal is in contact with the scala tympani, *via* the round window. Even if two holes are drilled, as in certain of Hallpike and Rawdon-Smith's cases, this relation usually obtains for anatomical reasons. Now it is known that the electrical response from the scala tympani is 180° out of phase with that from the scala vestibuli at low frequencies (Stevens, Davis and Lurie, 1935), and it may possibly be that the response from the VIII nerve, which is greater at low than at high frequencies as would be expected, thus tends to cancel the one effect and to augment the other. It is not impossible that this accounts for some of the differential effect demonstrated, and a similar criticism must apply to Stevens' and Davis' work on the guinea-pig, for there too the same spatial and electrical relations hold. It is not, however, likely that the whole effect is to be ascribed to this fact, and it may be taken that the relation shown constitutes strong evidence for the existence within the cochlea of some resonant structure.

What appears at first sight to be a further difficulty of this interpretation has recently arisen. Hallpike, Hartridge and Rawdon-Smith (1937 a, b) have shown that the responses from the cochlea and from the auditory tract exhibit quite dissimilar behaviour to a sudden phase-reversal in a continuous tone. The nervous response shows, as would be expected of impulses initiated by a resonant structure, an abrupt reduction immediately following the phase-change. The response is re-established to the tone in its new phase shortly thereafter. The cochlear response, however, shows no such interruption, and demonstrates a remarkable facility in following the changes of phase; it does so indeed at least as well as a piezo-electric sound-cell microphone, as these authors have shown. This fact argues against the origin of these potentials in a resonant structure. The observation on the neural response, however, simultaneously provides strong evidence for the resonance theory in that this response exhibits behaviour which can best be satisfactorily explained by assuming that the receptor cells of Corti's organ are supported on some resonant structure. Thus, whilst this latter fact greatly strengthens the resonance interpretation, the former argues against the validity of the earlier demonstration of a differential response. These apparent contradictions can, however, be explained if it is assumed, as was indicated earlier, that the cochlear response arises not from the basilar membrane or its associated structures themselves, but from some spatially related structure, such as the membrane of Reissner. Some independent evidence exists for this hypothesis, but into this it is not necessary to go here (Ashcroft, Hallpike and Rawdon-Smith, 1937; Gatty and Rawdon-Smith, 1937). Now it may be assumed that the response from the vibrating membrane of Reissner will be maximal over those regions where its amplitude of movement is maximal. In view of the presence of the cochlear fluids and of the extreme lightness of the membrane in question, this point will essentially be opposite the area of maximal vibration of the basilar membrane. Thus, to a steady note such as was employed in the differential response experiment, the voltages will be maximal at the base for high tones and at the apex for low. In the experiment involving changing the phase of the stimulus, however, the conditions are not altogether comparable. It may well be that the response of Reissner's membrane to a *transient* stimulus, such as the phase-change essentially constitutes, will be independent of the behaviour of the resonant basilar membrane. Moreover, the essential stimulus will reach the non-resonant membrane of Reissner, *via* the fluid of the scala vestibuli, slightly before it reaches the basilar membrane. It is necessary, therefore, to exercise some caution in interpreting the behaviour of these structures to transient stimuli in terms of data obtained from the study of their responses to continuous pure tones. Whether or not the explanation of these apparently discordant results given above is correct, however, only further experimental endeavour can decide.

It has been noted that the nervous response, which must be initiated in the hair cells of Corti's organ, behaves in a manner entirely different from the cochlear response, following a change of stimulus phase. This argues very strongly against Davis' theory of origin of the latter. According to this, the cochlear electrical response derives from the hair cells of the organ of Corti also (*e.g.* Davis, Derbyshire, Lurie and Saul, 1934). That the response is not characteristic of one generated by a resonant structure makes this supposition somewhat unlikely.

We may now attempt a summary of the evidence for and against the two main theories of pitch perception. In favour of the telephone theory is the important evidence that the frequency of a stimulus of relatively low pitch is reproduced in the form of synchronised nerve impulses in the auditory nervous system, at least in the midbrain. It is impossible, however, to detect any trace of such synchronised activity in the exposed cortex, except for stimuli of such low frequency that they cannot be confused with what are normally termed musical tones; but rather produce, to the human ear, a fluttering sensation. Against the telephone formulation is the strong evidence that the pitch of a note is not solely dependent upon its physical frequency. The resonance theory of pitch perception is supported by the demonstration that in all probability the cochlea does contain structures which constitute an adequate peripheral analytical mechanism and whose behaviour can most readily be interpreted on the assumption that, though heavily damped, they are none the less resonant.

It is not, of course, impossible that the differentiation of frequency which takes place along the basilar membrane occurs for some reason other than because this structure is differentially *tuned*. It may possibly be that it is not a system of tuned resonators, but instead that its differential response is due to the properties of fluids contained in tubes some part of whose walls or contents are elastic. Such an alternative to the conventional theory has been examined by Reboul (1938) and, although we have as yet no evidence for his view, it is equally true that there is no certain evidence for the opposing "resonance" one. Between these two possibilities as to the internal mechanics of the cochlea we cannot at present decide. We can at least say, however, that in default of evidence that any clue other than a spatial one is supplied to the cortex, a *place* theory must appear to be a sound basis for further speculation and experiment. An adequate theory of audition must attempt to explain, however, the second of the more obvious qualities of auditory experience, that of loudness or intensity. In the next chapter consideration may, therefore, be devoted to this matter.

CHAPTER 7

Theories of the Perception of Loudness

The data on intensity discrimination by the human ear are rather less complete than is the comparable matter on the discrimination of frequency. The first accurate work attempted here is that of Knudsen (1923), who used, as a source of sound, a thermionic valve audio oscillator, the output of which passed *via* a galvanometer and resistance network to a telephone installed in a sound-proof room. A motor driven key-switch operated to short-circuit one of the resistances of the network, and at this point the intensity of the sound in the telephone was increased. The amount of this intensity rise could be altered by adjustment of the value of the short-circuited resistance. Starting with an easily supraliminal change, this value was gradually decreased until the just noticeable difference was found. Such an apparatus has, however, very considerable technical defects. The use of an audio oscillator rather than a heterodyne oscillator restricted the frequency range of the investigation considerably, and is apt also to render the purity of the stimulus somewhat doubtful. Moreover, the introduction of an instantaneous intensity rise in the manner discussed carries with it certain important limitations. When either the absolute intensity or the intensity change or both are large, the arithmetical change of force on the diaphragm of the telephone in use at the moment of intensity change is sufficient to produce a series of damped oscillations at the natural frequency of the diaphragm. This phenomenon is objectionable, as it produces a readily audible click when the intensity change is effected. If Knudsen's technique is adopted, therefore, as the value of the intensity change is reduced so will the loudness of the click be reduced. A human subject of such an experiment must, therefore, make any judgement as to the perceptibility of the change in question at least partly on the basis of the loudness of the click accompanying the change. It is not possible to reduce this effect except by confining the investigation to notes of low intensity, an experimental procedure which would not yield all the required data, or by employing some other system for producing the intensity change.

Fortunately, Knudsen's work has been repeated by Riesz (1928) under more satisfactory conditions. The technique he employed was such that the deficiencies discussed above were avoided. It may here be stressed, however, that some confusion appears to have arisen in certain writings on this subject, as to the precise nature of Riesz' arrangements. Indeed, it has been stated that Knudsen's and Riesz' techniques were almost exactly similar (Banister, 1934, p. 889), although few similarities do, in fact, exist. This will at once be appreciated when it is realised that, in Reisz' apparatus no mechanical switch was employed, the required alterations in intensity being produced by feeding two oscillator generated voltages of slightly different frequency into a special telephone receiver. The magnitude of the beats thus produced could be adjusted by controlling the output of the two oscillators independently. Such a system, using as it does two sinusoidal inputs, produces changes of intensity which vary approximately sinusoidally with time, at a frequency corresponding to the difference in frequency of the outputs from the two oscillators. In this fashion all transient phenomena are avoided. The investigation was carried out at frequencies of 35, 70, 200, 1000, 4000, 7000, and 10,000 c.p.s., and at sensation levels of 5, 10, 20, 30, 40, 60, and 80 db. above the average human threshold. On these results certain observations may be made.

At every frequency tested, the Weber Fraction[1] is greater at the low than at the higher intensities. Moreover, the lower the stimulating frequency the greater is this departure from a constant relation between ΔI and I. Again, at the higher intensities (60–80 db. above threshold) $\Delta I/I$ is approximately constant over the frequency range tested, but at the lower intensities (5–20 db.

---

[1] *I.e.* $\Delta I/I$: I have adopted the term Weber Fraction in preference to the often used Weber Constant, because in no sense so far tested does Weber's Law hold over any but a most restricted range of absolute intensities, as we have noted above.

above threshold) $\Delta I/I$ is minimal at a frequency of about 2050 c.p.s. and becomes progressively much greater as the test frequency becomes lower or higher than this. Riesz has worked out a somewhat complicated formula to cover these results, and it is important to note that Weber's Law does not hold for the ear, except very approximately over the high intensity range; indeed, the Weber Fraction is, as Riesz shows, a continuous variable over the absolute intensity range investigated. Under optimal conditions, an intensity change of less than 10 per cent or approximately 0·4 db. can be detected, the Weber Fraction then being 1/10; under less than optimal conditions, a change of 73 per cent, or approximately 2·4 db., may be required (Riesz, 1928; Fletcher, 1929).

At this point a word of caution is in order. In contemporary works dealing with the subject of minimal perceptible differences it is customary to specify certain of the conditions under which the given threshold was measured. For light, the wave-length and brightness of the stimulus are customarily given; for sound, the frequency and intensity.[1] It is not usually realised, however, that the differential threshold value must depend upon many other variables, such as the rate of change of intensity at which the given measurement is made. This has been shown by the work of Rawdon-Smith and Grindley (1935), who have demonstrated that an intensity change, easily supraliminal if suddenly

---

[1] The term intensity, rather than the term loudness, is here used purposely. The brightness or brilliance of a light is normally measured by comparing the brightness of the unknown source with that of a source of known radiometric intensity. By varying the latter, a brilliance match is obtained, and thus a *photometric* value is reached; this gives an indication of the magnitude of the stimulus in terms of its power to stimulate. In sound, this procedure is rarely followed. It is customary to specify the intensity of a stimulus in terms of the ratio of its physical intensity to the intensity of a sound of similar frequency and wave form which is of loudness such that it is only just perceptible. Such a value bears little relation to the loudness level of a sound, which must be defined in a similar way to the brightness of a light source. Thus the loudness level of a sound may be measured only by comparing it with a sound of known frequency and waveform, whose intensity may be varied until a loudness match is secured. Of late an attempt has been made to establish such a system for measurements of loudness level, by comparing the unknown source with a pure tone of frequency 1000 c.p.s. and variable intensity. In this way, the loudness level may be measured in "phons" (British Standards Institution, 1936).

effected, may be subliminal if brought about slowly. From their results it is clearly evident that, in comparing sensory threshold values obtained by various methods, caution should be exercised to avoid a comparison of values secured under incomparable conditions.

It is now our task to examine the proposed theoretical mechanisms of loudness perception. On the classical resonance theory propounded by Helmholtz, intensity was supposedly perceived on the basis of the amplitude of vibration of that part of the basilar membrane excited by the stimulus; in modern terms, this might be correlated with the frequency of impulses in the sensory nerve attached to the appropriate hair cells of the organ of Corti. But it has already been pointed out that frequency of discharge is to be correlated, not with intensity, but with frequency of stimulus. On this point, then, Helmholtz cannot at first sight be thought to be correct. But careful analysis reveals a fault in this argument. Although the frequency of the electrical response in the whole of the VIII nerve may be correlated with stimulus frequency, it is by no means impossible that frequency of response in any individual nerve fibre may be correlated with intensity. And for this reason; assuming that the "volley" mechanism is operating, and that a single fibre is responding to every third stimulus wave, an increase of stimulus intensity may cause it to respond to every second wave; for, with increased amplitude of stimulation of the associated hair cell, the nerve fibre may now respond somewhat earlier in its *relative* refractory phase than was previously the case. Thus, though the frequency of the response in the whole VIII trunk may remain unchanged with a rise of stimulus intensity, the frequency of discharge in one single fibre may well have increased. Such a mechanism would probably provide one necessary sensory clue as to the loudness of the stimulus.

But another and additional mechanism may also be supposed, according to the "telephone" theory. When the intensity of the stimulus is increased, not only does the amplitude of vibration of the basilar membrane rise but so also does the *extent* of this vibration. The fibres of the membrane, connected as they are along their lengths in the manner previously discussed, will necessarily behave in the following way: with a rise in stimulus intensity, those fibres on either side of the few which, by virtue of their resonant period, respond most readily at the frequency of the stimulus, will also be forced to vibrate; this they may do with an amplitude sufficient to excite the hair cells of their associated organ of Corti. Thus we may imagine that, with a gradual increase in stimulus intensity, the area of basilar membrane vibrating will also gradually increase. The centre of this area will remain almost unchanged,[1] but its extent will become greater as the loudness of the stimulus becomes greater. This scheme need not necessarily belong only to the "telephone" theory, for, by the principle of maximal stimulation, which has been discussed earlier, the frequency of the stimulus will continue to be perceived on the basis of the spatial position of the centre of the resonant area; but at higher intensities, the number of active hair cells, and therefore the number of active nerve fibres, must also be higher. Thus, the second theoretical clue as to the loudness of a stimulus is, in brief, the number of conducting fibres; the greater this, the greater the perceived loudness.

Such hypothetical considerations are to some extent borne out by the relevant experimental investigations. It now becomes necessary, therefore, to consider certain of these findings.

Davis (1935) and Derbyshire and Davis (1935) show several curves from which it may be seen that the amplitude of the round window response does not bear a linear relation to the amplitude of the stimulus delivered to the tympanic membrane. The relationship shown is that between the logarithm of the stimulus intensity (decibels) and the amplitude of the electrical response; plotting the curves in this way, they are seen to be roughly sigmoid. At low stimulus intensities, the rise in amplitude is slow. Over the middle intensity range, the slope of the curve is greater, but this flattens out once more at high intensities. Now the round window response probably gives some indication of the amplitude of vibration of the basilar membrane. We should not expect, therefore, to obtain a linear relation between stimulus intensity and amplitude of response from the midbrain or VIII

[1] Though probably not absolutely unchanged (cf. p. 92).

nerve. But the relationship should be approximately linear between basilar membrane amplitude and amplitude of nervous electrical response. It is not likely that the rise in amplitude of the latter would express any quantal characteristic, for the number of nerve fibres potentially responding is so great that the addition of individual ones to those already active could not be seen as a discrete step, with any recording system yet available· These suppositions are amply borne out by the experimental data. The unequilibrated VIII nerve response is of almost the same relative amplitude as is that from the round window. It must be stressed that the partially or wholly equilibrated response, dependent as it is upon another variable, that of amount of equilibration at the stimulus frequency, does not show any such close correspondence.

There are two mechanisms whereby an increase in stimulus intensity may lead to an increase of response amplitude. These have already been discussed; either an increase in number of active fibres or an increase in frequency of response in any individual fibres will increase the voltage developed at the recording electrodes.[1] The evidence so far reviewed, conclusive as it is that the response increase is not quantal, does not differentiate between these two mechanisms. It may be accepted from general physical considerations that an increased stimulus intensity will increase the area of the basilar membrane vibrating. An increased number of active fibres must inevitably follow, though there is no relevant experimental evidence here. The demonstration of an increased frequency of response in any single fibre can however be made more directly.

We have already noted that when stimulating at high frequencies, the synchrony of the nervous response is found to have broken down. The electrical response is, in fact, merely a series of random transients, corresponding to the passage of random and unsynchronised nervous impulses. For all frequencies above some 3000 c.p.s., no synchronised response can be found, no matter how great the stimulus intensity. But at frequencies

---

[1] An increase in amplitude from the latter cause will of course only be manifested up to a point where all the responding fibres show complete synchrony; i.e. until every fibre is responding to every stimulus cycle.

between 2000 and 2500 c.p.s., an interesting effect may be observed. Whilst for a stimulus of low intensity the response is unsynchronised, if the stimulus intensity is increased the response becomes partially synchronised; this must be interpreted as meaning that, for the low intensity, the stimulus is insufficient to excite the nerve fibre in its partially refractory period. With a greater stimulus, however, the fibre responds rather more regularly, for the increased stimulus is sufficient to excite it in a state of submaximal sensitivity. When this happens in a number of fibres, there may, at the higher intensity, be sufficient fibres responding to a single stimulus wave to reconstitute this in the electrical response, whilst at the lower intensity too few fibres were responding to any single wave to render this possible (author's unpublished observation).

In summary, then, it may be said that there is evidence that variations in loudness may be correlated both with variations in the number of active fibres in the VIII nerve and midbrain, and with frequency of impulses in any single fibre; the latter supposition is, it may be noted, in accordance with the classical data provided by Adrian and others in their work on other sense organs (Adrian, 1932).

No theoretical treatment of this subject can make any claim to completeness unless it devote careful consideration to the various objections which Davis and his co-workers have advanced against the inclusion of the latter supposition; i.e. that loudness is to be correlated with the frequency of discharge in individual fibres. Its adoption for the sole reason that this phenomenon is general in other sense organs is insupportable. It must now be our purpose, therefore, to review Davis' objections in the light of the relevant experimental data.

It has been pointed out above that the response from the VIII nerve or the auditory midbrain demonstrates a phenomenon known as equilibration, whereby the amplitude of the response shows, for a fraction of a second after the onset of a stimulating tone, a shrinkage in size. This effect is greatest at frequencies near to those at which the phenomenon of alternation takes place (*vide supra*), that is at frequencies which, it may be supposed, are multiples of the reciprocal of the average absolute refractory phase of the fibres of the nerve in question. In a good preparation, these frequencies are multiples of 800–1000 c.p.s. which gives a value for the refractory period of 1–1·2 ms. (Davis, 1935; Derbyshire and Davis, 1935). The interpretation of the alternation effect is simple enough; at frequencies which, as has been said, are multiples of the reciprocal of the refractory period, the initial response amplitude (excluding the "on-effect") is approximately one-half to one-third the response to a frequency immediately below these "critical" frequencies. This is because, above a critical frequency no one fibre can respond to the stimulus at the same rate as it can to a slightly lower frequency. It is forced to respond less often, with the result that the amplitude of the electrical response from the whole nerve falls. The electrical response amplitude is dependent upon the number of fibres contributing to each single response cycle, and, as the fibres now respond alternately, only half the number activated for each response wave at the lower frequency are active at the higher. Now at the next critical frequency, 2000 c.p.s., the fibres which below this frequency respond in the 1 : 2 relationship just outlined must now respond to every *third* wave; effectively, then, one-third the maximum number of fibres will be operative for each response wave. The electrical response should thus have fallen to one-third that at frequencies below 1000 c.p.s.

From this and the preceding discussion (p. 99) it will be seen that the phenomenon of alternation is sufficiently well established for its acceptance to be warranted; moreover, the theoretical basis advanced by Davis and his assistants would appear to be perfectly tenable. Such of his theory as we have so far expounded makes clear, also, the further hypothetical step whereby one may account for the process of equilibration. As was stated above, it is necessary only to suppose that, like other nerve fibres (Field and Brücke, 1926), those of the VIII nerve suffer a prolongation of relative refractory period as a result of prolonged stimulation. As the refractory period lengthens, so does the number of fibres operative for each response wave fall. Another and secondary effect may also be supposed to take place. As was shown by Forbes and Rice (1929), the action potential of a nerve fibre shows a decrease in amplitude as stimulation is continued; this is particularly the case if the fibre is stimulated at a frequency approximating to that which, by virtue of its refractory phase, is the fastest it may transmit. Thus, "equilibration" in nerve trunks consists of a reduction in amplitude of the electrical response due to (a) failure of some fibres to respond owing to an increase in relative refractory phase, and (b) a decreased action potential in individual fibres (cf. also Gerard and Marshall, 1933).

It is clear that, if loudness is to be perceived at least partly on the basis of frequency of discharge in individual nerve fibres, it might be expected that these physiological changes found in the cat would correspond to a subjective change in the human being, when listening to sustained tones. This is, indeed, Davis' first argument against the inclusion of this mechanism in a theory of loudness perception (Davis, 1935). "We might expect", he says (p. 209, col. 2), "a greater susceptibility to fatigue corresponding to the equilibration, or an abrupt change in loudness with shift of frequency near the critical points. These have not been reported in studies of fatigue or loudness, and we must conclude that our judgement of loudness depends not upon the number of impulses per second arriving at the auditory nerve centres but instead upon the number of fibres which are delivering impulses, irrespective of their frequency."

At first sight this argument must seem fairly conclusive; careful examination, however, reveals certain fallacies. Thus, the fact that no sudden change of loudness takes place at the critical frequencies is not surprising, for at least two reasons. First, the amplitude decrease at 1000 c.p.s. or thereabouts is only 6 db. at the most. Moreover, it is usually fairly slow; that is to say, the full decrease in amplitude is not manifested in many preparations, until a frequency of 1800 cycles is reached (Derbyshire and Davis, 1935, p. 479, and *vide* Fig. 16 (*a*) of the present work). This is equivalent to saying, therefore, that if the frequency of a pure tone stimulus is steadily increased from (say) 900 c.p.s. to 2000 c.p.s. an intensity change of 6 db. being simultaneously introduced, this latter should be readily detected; this despite the change taking place during an appreciable period of time, perhaps 1 or 2 sec. There is every reason to suppose that this would not be so, for such a change, accomplished at such a rate, would almost certainly be subliminal.

Another point may be made in this connection. In the paper quoted above, Derbyshire and Davis (1935) note (p. 501) that "There is no suggestion of any auditory experience that correlates with the particularly rapid and extensive equilibration which occurs in the auditory nerve of the cat at frequencies near 1000 and 2000 c.p.s." The most striking case of equilibration is shown by Davis (1935); the amplitude decrease amounts to 8 db. in 50 ms. At least some of this is due, not to decreasing frequency of response in individual fibres, but to decreased electrical manifestation, as pointed out above. Some of the loss will be due to the fact that the nerve fibres attached to those hair cells at the ends of the vibrating strip of basilar membrane are no longer stimulated above threshold. The remainder will be due to increased refractory period. Thus we may assume that the subjective loudness loss would amount to considerably less than 8 db. Moreover, this change would take place once only, at the beginning of the tone, and would complete in 50 ms. It is scarcely necessary to point out that, under these conditions, this change would almost certainly be unnoticed. In any case, too, the organism would have become quite habituated to this series of events.

Secondly, with regard to Davis' contention that the ear should be more susceptible to fatigue at frequencies near to the critical ones, it should be pointed out that this hypothesis necessarily involves that the phenomenon of "auditory fatigue" should consist in a loss of sensitivity incurred by some change in the peripheral mechanism, the VIII nerve and auditory midbrain. It has been very clearly shown, however, that this is not, in fact, the case; auditory fatigue appears in part to be a phenomenon of cortical mediation (Rawdon-Smith, 1934, 1936), and it is for this reason that Davis' statement here, also, cannot be accepted. It is probably true that there is also a fatigue effect of the type which he suspects, of peripheral locus; but any investigation of this alone, on the human subject, is rendered impossible for the reason that these relatively small changes in auditory sensitivity are often masked by much larger changes of central

(a)

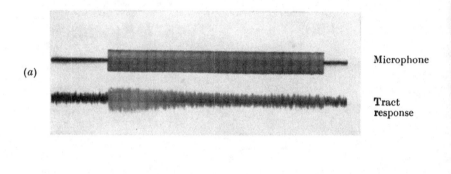

Microphone

Tract
response

(b)

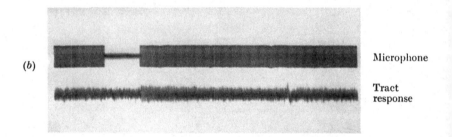

Microphone

Tract
response

Fig. 18. (a) Oscillogram showing reduction in amplitude of nervous response during continuous stimulation for *one minute*. Upper record
shows constancy of sound stimulus, as measured by microphone.
Frequency of stimulus: 800 c.p.s. Intensity of stimulus: approx. 50
microwatts/cm². (b) Oscillogram showing incomplete recovery of
sensitivity after a rest period of *ten seconds*. For initial amplitude of
response, see (a). Upper record from microphone. Frequency and
intensity of stimulus as in (a).

mediation. We may, however, summarise the evidence in regard to the peripheral mechanism.

The equilibration effect which we have discussed takes place in about 2 sec. from the onset of the stimulating tone (Derbyshire and Davis, 1935, p. 481). There is, however, a second and, from our point of view, much more important change which manifests itself, after the equilibration effect is over, for very intense stimuli. From the records shown in Fig. 18 (*a*), it may be clearly seen that the amplitude of the electrical response from the midbrain does not always settle down after the initial equilibration effect is over, but for such stimulating conditions may continue to fall. This effect appears to be a true fatigue one, for, as may be seen from Fig. 18 (*b*), the initial amplitude of discharge is not restored after a rest period; the length of this latter may be between 5 and 30 sec., and in all cases the degree of recovery depends on this length of time. The longer the rest, the greater the recovery.

The record in Fig. 18 (*a*) shows the result of one minute of stimulation. The equilibration effect proper is over in some two seconds, as has been pointed out above; this would, therefore, correspond to the first 2·5 mm. of the record shown. By careful choice of the stimulus frequency, however, equilibration is almost negligible in this case. Thus, the total amplitude drop, approximating to 6 db., may be considered to be due to an effect to be distinguished from equilibration; for the purposes of this work, it is proposed to designate this latter *fatigue*.

The fatigue effect has been found for all frequencies tested between 400 and 1000 c.p.s. The latter figure represents the highest stimulus frequency which will elicit a satisfactory response from the contralateral *inferior colliculi* in the cat. Below 400 c.p.s. a trace of fatigue may sometimes be detected, but this is in all cases too small to be considered significant.

Derbyshire and Davis (1935) have shown an effect precisely similar to fatigue in the VIII nerve response. This they designate "Slow Equilibration", but there is every reason to suppose that it corresponds exactly to the phenomenon we have found. As it represents, moreover, the physiological aspect of a phenomenon which is subjectively demonstrable (Rawdon-Smith, 1934,

1936), and as it has all the characteristics of a fatigue effect, this latter terminology would, it is thought, appear preferable.

In parenthesis it must be said that the fatigue effect here described might possibly be due to some increase of stiffness produced by contraction of the two muscles of the middle ear, namely the *tensor tympani* and *stapedius*. If, after sustained stimulation of the ear, the tension exerted by these muscles was increased, it might be expected that the amplitude of oscillation of the middle ear members would be decreased, in virtue of the increase of stiffness produced in this way. But this possibility may be excluded in the following manner. By recording the amplitude of the cochlear effect, a measure is obtained of the magnitude of vibration of the inner ear members. With increased contraction of the muscles referred to, this amplitude decreases, as we have noted above (p. 73). Immediately on switching off the stimulus, however, the normal tonic contraction of the muscles is restored, and no effect persists from one stimulus to another, so far as may be ascertained. Thus no "fatigue" phenomena can be ascribed to such factors.

It is worthy of note that Békésy (1929) has shown that fatigue may, in the human subject, cause erroneous judgements of pitch. Thus he has shown that a fatiguing note of 800 c.p.s. causes a sharpening of pitch for tones from 800 to 1500 c.p.s., the greatest effect being about one semitone (actually 7 per cent) at 1200 c.p.s. For tones below 800 c.p.s., that is from 800 to 300 c.p.s., a similar flattening is demonstrable. This is greatest at 500 c.p.s., these being approximately equal to a frequency shift of 6 per cent. A somewhat similar result has been reported by Cathcart and Dawson (1929), and, in the opinion of Beatty (1932, p. 116) may be similarly interpreted.

This effect is considered by Békésy to constitute strong proof of the resonance hypothesis of frequency perception. He has (1929) elaborated a theory wherein it is supposed that the result of fatiguing the ear is to produce an apparent change in the "point of maximal stimulation" considered earlier in this work. As the receptors attached to the true point of maximal stimulation are fatigued, a maximum nervous response is obtained, not from the point of maximal stimulation, but from a less fatigued area nearby. As a result, the judgement of the observer as to pitch is slightly upset, as may be readily understood. But thus the hypothesis involves that pitch is judged not on the basis of the *geometrical* centre of the vibrating area, but as the position corresponding to those fibres of the VIII nerve transmitting a maximal number of impulses in unit time. If this is so, and Békésy's theory is attractive, this work must be considered to constitute further evidence against Davis' rejection of the theory that *loudness* is judged, at least partly, on the same basis. For it is clear that, if frequency of nervous discharge is to be represented in the auditory cortex for the purpose of permitting judgement of the spatial position of a maximally stimulated set of end-organs, and if this frequency of discharge is, as it must be, also a function of intensity of stimulation, then there is every reason to suppose that judgements of the latter will be made partly on the basis of the former.

Such data as are available on the general phenomenon of auditory fatigue, therefore, offer no support for Davis' rejection of the hypothesis that loudness is judged, even if only partly, on the basis of the frequency of impulses in individual fibres of the acoustic nerve. Our final picture of the mechanism of intensity discrimination in the human ear, is, therefore, a dual one. On the one hand, as we have seen, the number of nerve fibres active will depend on the degree of spread of the vibrating area of the basilar membrane, and this in turn on the intensity of the stimulus. On the other hand, there is evidence that the frequency of impulses in the conducting fibres will also be governed, to some extent, by the intensity of the incoming stimulus. We must suppose, therefore, that the frequency of impulses will provide a second clue for our judgements of loudness.

POSTSCRIPT

In the preceding chapters we have seen how it is thought the eye and ear operate to enable us to discriminate differences of amplitude and frequency in light and sound. Such discrimination, we believe, is fundamental to our visual and auditory experience, and a full understanding of the mode of operation of our two most important sense organs in these respects will tell us much of the way in which even our more complicated sensations arise. I have attempted, in this book, to give the reader some idea of the more important approaches to this problem, and to indicate how far we have explored them at the present time. To what extent the predictions I have made here will be justified it is impossible at present to say. But a steady evolution of new techniques and a rapid accumulation of new ideas has characterised the last decade, and, should these advances be maintained, we may hope to clear up the obscurities remaining with reasonable rapidity; in this way to open another road to the fundamental problem of central nervous physiology itself.

It is my hope that this book may encourage others to pursue the course whose direction has been so clearly indicated by Newton, Young and von Helmholtz, those far-sighted pioneers in this important field.

BIBLIOGRAPHY

SECTION I: Vision

ABNEY, W. de W., 1913. *Researches in Colour Vision.* London.

ADRIAN, E. D., 1928. *The Basis of Sensation.* London: Christophers.

—— 1932. *The Mechanism of Nervous Action,* pp. 2–19. Philadelphia: Univ. of Pennsylvania Press; London: Oxford University Press.

ADRIAN, E. D. and BRONK, D. W., 1928. *J. Physiol.* **66**, 81.

ADRIAN, E. D. and MATTHEWS, R., 1927a. *J. Physiol.* **63**, 378.

—— —— 1927b. *J. Physiol.* **64**, 279.

—— —— 1928. *J. Physiol.* **65**, 273.

ADRIAN, E. D. and ZOTTERMAN, Y., 1926. *J. Physiol.* **61**, 151.

ALLEN, F., 1919. *Phil. Mag.* **38**, 81.

—— 1926. *J. Opt. Soc. Amer.* **13**, 383.

AREY, L. B., 1915a. *J. Comp. Neurol.* **2**, 535.

—— 1915b. *Science,* **42**, 915.

BAKER, T. Y. and BRYAN, G. B., 1912. *Proc. Optical Convention,* p. 253.

BLANCHARD, J., 1918. *Phys. Rev.* **11**, 18.

BOLL, F., 1876. *Monatsber. Berlin. Akad.* p. 783.

BOUGUER, P., 1760. *Traité d'optique sur la gradation de la lumière.* Paris.

BROCA, A., 1901. *J. de Physiol. et de Path. gén.* **3**, 384.

BRONK, D. W., 1929. *J. Physiol.* **67**, 17.

CRAIK, K. J. W., 1938. *J. Physiol.* **92**, 406.

CRAWFORD, B. H., 1937. *Proc. Roy. Soc. B,* **124**, 81.

DEMOLL, R., *Zool. Jahrb.* **38**, 441.

DIETER, W., 1929. *Pflüger's Arch.* **222**, 381.

FECHNER, G. T., 1860. *Elemente der Psychophysik.* Leipzig.

GIBSON, K. S. and TYNDALL, E. P. T., 1923. *Sci. Papers Bur. Stand.* **19**, No. 475.

GUILD, J., 1931. *Phil. Trans. Roy. Soc.* **230**A, 149.

—— 1932. *Report of a Joint Discussion on Vision,* p. 167. Phys. Soc., London.

HARTLINE, H. K. and GRAHAM, C. H., 1932. *J. Cell. and Comp. Physiol.* **1**, 277.

HARTRIDGE, H., 1918. *J. Physiol.* **52**, 175.

—— 1923. *J. Physiol.* **57**, 52.

HECHT, S., 1921. *J. Gen. Physiol.* **4**, 113.

—— 1924. *J. Gen. Physiol.* **7**, 235.

HECHT, S. 1928 a. J. Gen. Physiol. 11, 255.

—— 1928 b. Proc. Nat. Acad. Sci. (Washington), 14, 237.

HECHT, S., 1930. J. Opt. Soc. Amer. 20, 231.

—— 1934. Vision II in Handbook of General Experimental Psychology (edited by C. Murchison), pp. 704–828. Worcester: Clark University Press; London: Oxford University Press.

—— 1935. J. Gen. Physiol. 18, 767.

—— 1937. Physiol. Rev. 17, 2, 239.

HECHT, S., HAIG, C. and CHASE, M., 1937. J. Gen. Physiol. 20, 6, 831.

HECHT, S. and WILLIAMS, R., 1922. J. Gen. Physiol. 5, 1.

VON HELMHOLTZ, H., 1866. Handbuch der physiologischen Optik, 1st edition. Hamburg and Leipzig: Voss.

—— 1891. Ztsch. f. Sinnesphysiol. 3, 517.

—— 1909–11. Handbuch der physiologischen Optik, 3rd edition. Hamburg and Leipzig: Voss.

—— 1924. Physiological Optics (edited by J. P. C. Southall). New York: Optical Society of America.

HERING, E., 1888. Vorgänge der lebenden Materie: Prague.

—— 1907. Grundzüge der Lehre vom Lichtsinn in Graefe-Saemisch Handbuch der Augenheilk. 3, Kap. 12. Berlin.

—— 1920. Grundzüge der Lehre vom Lichtsinn in Graefe-Saemisch Handbuch der Augenheilk. 1, Kap. 13, 213. Berlin.

HOFMANN, F. B., 1920. Die Lehre vom Raumsinn des Auges in Graefe-Saemisch Handbuch der Augenheilk. 1, Kap. 13, 213. Berlin.

HOUSTOUN, R. A., 1932 a. Vision and Colour Vision. London: Longmans.

—— 1932 b. Report of a Joint Discussion on Vision, p. 167. London: Physical Society.

IVES, H. E., 1912. Phil. Mag. 24, 352.

KOHLRAUSCH, A., 1922. Pflüger's Arch. 196, 113.

—— 1931. Handbuch der normalen und pathologischen Physiologie (edited by A. Bethe, et al.), 12, Pt. 2, 1499. Berlin: Springer.

KÖNIG, A., 1897. Sitzungsb. Akad. Wissensch. Berlin, p. 559.

—— 1903. Gesammelte Abhandlungen zur physiologischen Optik. Leipzig: Barth.

KÖNIG, A. and BRODHUN, E., 1888. Sitzungsb. Akad. Wissensch. Berlin, p. 917.

—— —— 1889. Sitzungsb. Akad. Wissensch. Berlin, p. 641.

KÖNIG, A. and DIETERICI, C., 1892. Ztsch. f. Psychol. 4, 241.

VON KRIES, J., 1895. Ztsch. f. Psychol. u. Physiol. d. Sinnesorgane, 9, 81.

—— 1905. Physiologie der Sinne (edited by W. Nagel) in Handbuch der Physiologie des Menschens. Braunschweig: Vieweg.

—— 1929. Handbuch der normalen und pathologischen Physiologie (edited by A. Bethe, et al.), 13, 678. Berlin: Springer.

KÜHNE, W., 1879. Handbuch der Physiologie (edited by L. Hermann), 3, Pt. 1, 235. Leipzig: Vogel.

KUNDT, A., 1878. Ann. Phys. u. Chem. 4, 34.

LADD-FRANKLIN, C., 1893. *Mind*, N.S., **2**, 473.
—— 1929. *Colour and Colour Theories*. New York: Harcourt.
LASAREFF, P., 1923. *Pflüger's Arch.* **199**, 290.
LUDVIGH, E. J., 1938, *Arch. Ophth.* (in press).
LYTHGOE, R. J., 1932. *Medical Research Council, Spec. Rep. Series*, No. 173.
—— 1937, *J. Physiol.* **89**, 331.
LYTHGOE, R. J. and TANSLEY, K., 1929 a. *Medical Research Council, Spec. Rep. Series*, No. 134.
—— —— 1929 b. *Proc. Roy. Soc.* B, **105**, 60.
McDOUGALL, W., 1901. *Mind*, N.S., **10**, 52, 210, 347.
MATTHEWS, B. H. C., 1931. *J. Physiol.* **71**, 64.
MAXWELL, J. C., 1855, *Trans. Roy. Scot. Soc. Arts*, **4**, Pt. 2.
—— 1890. *Scientific Papers* (edited by W. D. Niven), Vol. 1. London: Cambridge University Press.
MÜLLER, G. E., 1896. *Ztsch. f. Psychol. u. Physiol. d. Sinnesorgane*, **10**, 321, and **14**, 161.
—— 1924. *Typen der Farbenblindheit, u.s.w.*, Göttingen Nachrichten.
MYERS, C. S., 1928. *A Text Book of Experimental Psychology*, 3rd edition, Part 1, p. 247. Cambridge University Press.
NEWTON, I., 1794. *Opticks* (4th edition reprinted 1931). London: Bell.
PARINAUD, H., 1881. *Compt. rend. Acad. Sci.* **93**, 286.
PARSONS, J. H., 1924. *An Introduction to the Study of Colour Vision*. Cambridge University Press.
PIPER, H., 1903. *Ztsch. f. Psychol. u. Physiol. d. Sinnesorgane*, **31**, 161.
PRIEST, I. G., 1918. *Phys. Rev.* **11**, 502.
PRIEST, I. G. and BRICKWEDDE, F. G., 1926. *J. Opt. Soc. Amer.* **13**, 306.
PURDY, D. M., 1935. *Vision*: Chapter 4 of *Psychology*, by Boring, E. G. *et al.* pp. 57–101.
QUAIN, T., 1909. *Anatomy* (edited by E. A. Schäfer, *et al.*), Vol. 3, Pt. II. London and New York: Longmans.
RIVERS, W. H. R., 1900. *Vision*, in *Schäfer's Textbook of Physiology*, **2**, 1026.
ROCHAT, G. F., 1925. *Arch. néerl. de physiol.* **10**, 448.
ROELOFS, C. O. and ZEEMAN, W. P. C., 1919. *Arch. Ophth.*, Leipzig, **99**, 174.
SINDEN, R. H., 1923. *J. Opt. Soc. Amer.* **7**, 1123.
SMITH, F. O., 1925. *J. Exper. Psychol.* **8**, 381.
SMITH, T. R., 1936. *J. Gen. Psychol.* **14**, 318.
STEINDLER, O., 1906. *Sitzber. d. Wien. Akad. Math.-naturwiss.* **115**, Abt. 2 a, 39.
STEINHARDT, T., 1936. *J. Gen. Physiol.* **20**, 185.
STILES, W. S., 1937. *Proc. Roy. Soc.* B, **123**, 90.
STILES, W. S. and CRAWFORD, B. H., 1933. *Proc. Roy. Soc.* B, **112**, 428.

TANSLEY, K., 1931. *J. Physiol.* 71, 442.

TRENDELENBURG, W., 1904. *Ztsch. f. Psychol. u. Physiol. d. Sinnesorgane*, 37, 1.

—— 1911. *Ergebn. Physiol.* 11, 1.

—— 1923. *Pflüger's Arch.* 201, 235.

TROLAND, L. T., 1922. *J. Opt. Soc. Amer.* 6, 531.

—— 1934. *Vision I* in *Handbook of General Experimental Psychology* (edited by C. Murchison), pp. 653–703. Worcester: Clark University Press; London: Oxford University Press.

WEBER, E. H., 1835. *Arch. Anat. u. Physiol.* 152.

—— 1846. *Der Tastsinn u. d. Gemeingefühl* in Wagner's *Handwörterbuch d. Physiol.* 3, Pt. 2, 481. Berlin: Braunschweig.

WEIGERT, F., 1921. *Pflüger's Arch.* 190, 177.

WEISS, P., 1931. *Pflüger's Arch.* 226, 600.

WESTPHAL, H., 1909. *Ztsch. f. Sinnesphysiol.* 44, 182.

WILCOX, W. W., 1932. *Proc. Nat. Acad. Sci.* 18, 47.

WRIGHT, W. D., 1928. *Trans. Opt. Soc.* 30, 141.

—— 1929. *Trans. Opt. Soc.* 31, 201.

—— 1934. *Proc. Roy. Soc.* B, 115, 49.

SECTION II: Audition

ADES, H. W., METTLER, F. A. and CULLER, E. A., 1938. *Proc. Amer. Physiol. Soc.* 31 March, 1938. *Amer. J. Physiol.* (in press).

ADRIAN, E. D., 1928. *The Basis of Sensation.* London: Christophers.

—— 1931a. *J. Physiol.* 71, 28 P.

—— 1931b. *Discussion on Audition*, p. 5. London: Physical Society.

—— 1932. *The Mechanism of Nervous Action*, Philadelphia: Univ. of Pennsylvania Press; London: Oxford Univ. Press.

ASHCROFT, D. W. and HALLPIKE, C. S., 1934. *J. Laryngol. and Otol.* 49, 450.

ASHCROFT, D. W., HALLPIKE, C. S. and RAWDON-SMITH, A. F., 1937. *Proc. Roy. Soc.* B, 122, 186.

BANISTER, H., 1926a. *Phil. Mag.* 2, 402.

—— 1926b. *Brit. J. Psychol.* 16, 265.

—— 1926c. *Brit. J. Psychol.* 17, 142.

—— 1934, *Audition I* in *Handbook of General Experimental Psychology* (edited by C. Murchison), pp. 880–923. Worcester: Clark University Press; London: Oxford University Press.

BARTON, E. H., 1922. *Textbook of Sound.* London: Macmillan.

BEATTY, R. T., 1932. *Hearing in Man and Animals.* London: Bell.

VON BÉKÉSY, G., 1929. *Phys. Ztsch.* 30, 115.

BORING, E. G., 1926. *Amer. J. Psychol.* 37, 157.

BRITISH STANDARDS INSTITUTION, 1936. *Glossary of Acoustical Terms and Definitions.* London: British Standards Institution.

CATHCART, E. P. and DAWSON, S., 1929. *Brit. J. Psychol.* 19, 343.
COSENS, C. R. G. and HARTRIDGE, H., 1922. *Brit. J. Psychol.* 13, 48.
CROWE, S. J., GUILD, S. R. and POLVOGT, L. M., 1934. *Bull. Johns Hopkins Hosp.* 54, 31.
DAVIS, H., 1934. *Audition III* in *Handbook of General Experimental Psychology* (edited by C. Murchison), pp. 962–86. Worcester: Clark University Press; London: Oxford University Press.
—— 1935. *J. Acoust. Soc. Amer.* 6, 205.
DAVIS, H., DERBYSHIRE, A. J., KEMP, E. H., LURIE, M. H. and UPTON, M., 1935. *J. Gen. Psychol.* 12, 251.
DAVIS, H., DERBYSHIRE, A. J., LURIE, M. H. and SAUL, L. J., 1934. *Amer. J. Physiol.* 107, 311.
DAVIS, H., FORBES, A. and DERBYSHIRE, A. J., 1933. *Science*, 78, 522.
DAVIS, H., LURIE, M. H. and STEVENS, S. S., 1935. *Ann. Otol. Rhinol. and Laryngol.* 44, 776.
DAVIS, H. and SAUL, L. J., 1931. *Science*, 74, 205.
—— —— 1932. *Amer. J. Physiol.* 101, 28.
DERBYSHIRE, A. J. and DAVIS, H., 1935. *Amer. J. Physiol.* 113, 476.
DUSSER DE BARENNE, T. G., 1934. *The Labyrinthine and Postural Mechanisms*, in *Handbook of General Experimental Psychology* (edited by C. Murchison), pp. 204–46. Worcester: Clark University Press; London: Oxford University Press.
EWALD, J. R., 1903. *Pflüger's Arch.* 93, 485.
EWALD, J. R. and JÄDERHOLM, O. A., 1908. *Pflüger's Arch.* 124, 29.
FIELD, H. and BRÜCKE, E. T., 1926. *Pflüger's Arch.* 214, 103.
FLETCHER, H., 1929. *Speech and Hearing*. New York: van Nostrand; London: Macmillan.
—— 1930. *J. Acoust. Soc. Amer.* 1, 311.
FORBES, A. and RICE, L., 1929. *Amer. J. Physiol.* 90, 119.
GASSER, H. S. and ERLANGER, J., 1925. *Amer. J. Physiol.* 73, 629.
GATTY, O. and RAWDON-SMITH, A. F., 1937. *Nature*, 139, 670.
GERARD, R. W. and MARSHALL, W. H., 1933. *Amer. J. Physiol.* 104, 575.
GRAY, A. A., 1900. *J. Anat. and Physiol.* 34, 324.
GUILD, S. R., 1932. *Acta Otolaryngol.* 17, 207.
—— 1935. *Ann. Otol. Rhinol. and Laryngol.* 44, 738.
GUILD, S. R., CROWE, S. J., BUNCH, C. C. and POLVOGT, L. M., 1931. *Acta Otolaryngol.* 15, 269.
HALLPIKE, C. S., HARTRIDGE, H. and RAWDON-SMITH, A. F., 1937a. *Proc. Physical Soc.* 49, 190.
—— —— —— 1937b. *Proc. Roy. Soc.* B, 122, 175.
HALLPIKE, C. S. and RAWDON-SMITH, A. F., 1934a. *J. Physiol.* 81, 25P.
—— —— 1934b. *Nature*, 113, 614.

126   SECTION II: AUDITION

HALLPIKE, C. S. and RAWDON-SMITH, A. F., 1934c. *J. Physiol.* 81, 395.
HARDESTY, I., 1908. *Amer. J. Anat.* 8, 104.
HARTRIDGE, H., 1921. *Brit. J. Psychol.* 12, 142.
—— 1922. *Brit. J. Psychol.* 12, 362.
—— 1934a. *Audition II* in *Handbook of General Experimental Psychology* (edited by C. Murchison), pp. 924–61. Worcester: Clark University Press; London: Oxford University Press.
—— 1934b. *J. Physiol.* 81, 15 P.
HELD, H. and KLEINKNECHT, F., 1927. *Pflüger's Arch.* 216, 1.
VON HELMHOLTZ, H., 1912. *On Sensations of Tone* (translated by A. J. Ellis). London: Longmans.
HORTON, G. P., 1933. *Psychol. Bull.* 31, 548.
—— 1934. *J. Comp. Psychol.* 18, 405.
KATO, T., 1913. *Pflüger's Arch.* 150, 569.
KEITH, A., 1918. Appendix to Wrightson (1918), *q.v.*
KEMP, E. H., 1935. *Psych. Bull.* 32, 5, 325.
KNUDSEN, V. O., 1923. *Phys. Rev.* 21, 1.
KÖHLER, W., 1909. *Ztsch. f. Psychol.* 54, 250.
LORENTE DE NÓ, R., 1933. *Laryngoscope*, 43, 1.
LUCIANI, L., 1917. *Human Physiology*, vol. 4, *The Sense Organs* (translated by F. A. Welby; edited by G. M. Holmes). London: Macmillan.
VAN DER POL, B., 1929. *Phil. Mag.* 7, 477.
POLIAK, S., 1932. *The Main Afferent Fiber Systems of the Cerebral Cortex in Primates.* Univ. Calif. Publ. Anat. 2.
RAWDON-SMITH, A. F., 1934. *Brit. J. Psychol.* 35, 78.
—— 1936. *Brit. J. Psychol.* 26, 233.
RAWDON-SMITH, A. F. and GRINDLEY, G. C., 1935. *Brit. J. Psychol.* 26, 191.
REBOUL, J. A., 1938. *J. de physique*, 9, Ser. 7.
RIESZ, R. R., 1928. *Phys. Rev.* 31, 867.
ROSS, D. A., 1936. *J. Physiol.* 86, 117.
SAUL, L. J. and DAVIS, H., 1932. *Arch. Neurol. and Psychiat.* 28, 1104.
SHOWER, E. G. and BIDDULPH, R., 1931. *J. Acoust. Soc. Amer.* 3, 275.
STEVENS, S. S., 1935. *J. Acoust. Soc. Amer.* 6, 150.
STEVENS, S. S., DAVIS, H. and LURIE, M. H., 1935. *J. Gen. Psychol.* 13, 297.
STEVENS, S. S. and NEWMAN, E. B., 1936. *Amer. J. Psychol.* 48, 297.
TROLAND, L. T., 1929. *J. Gen. Psychol.* 2, 28.
—— 1930. *The Principles of Psychophysiology*, 2, 222. New York: van Nostrand.
WALKER, A. E., 1937, *J. Anat.* 71, 319.
WEVER, E. G., 1933. *Physiol. Rev.* 13, 3.

WEVER, E. G. and BRAY, C. W., 1930 a. *Science*, 71, 215.

WEVER, E. G. and BRAY, C. W., 1930 b. *Proc. Nat. Acad. Sci. (Washington)*, **16**, 344.

—— —— 1930 c. *J. Exp. Psychol.* **13**, 373.

—— —— 1930 d. *Psychol. Rev.* **37**, 5, 365.

—— —— 1936. *J. Psychol.* **3**, 101.

WEVER, E. G., BRAY, C. W. and HORTON, G. P., 1934. *Science*, **80**, 18.

WILKINSON, G. and GRAY, A. A., 1924. *The Mechanism of the Cochlea*. London: Macmillan.

WITTMAACK, K., 1907. *Ztsch. f. Ohrenheilk.* **54**, 37.

WRIGHTSON, T., 1918. *An Inquiry into the Analytical Mechanism of the Internal Ear*. London: Macmillan.

YOSHII, V., 1909. *Ztsch. f. Ohrenheilk.* **58**, 201.

ZURMÜHL, G., 1930. *Ztsch. f. Sinnesphysiol.* **61**, 40.

INDEX

Printed in the United States
By Bookmasters